Building the Case: How Lawyers are Leading on Sustainability

Artist Statement: The authors of this publication are all grappling with what sustainability means to them. Personally, professionally, ethically, they are all reaching out a hand, trying to find the shape of a future that can support its own weight—the balance of industry and resource, the cost of both remediating and innovating. The values of the individual are writ large on the broader landscape of our society, and how we compromise in order to build a lasting tomorrow. Two hands reach toward each other, almost ready to grasp, to shake—overlapping but not quite solid in separation. The colors of a binary star: blue and red, the inexorable pull of gravity between them, the necessity of their endless balance. Surrounded by a golden glow—hopeful and blinding—shaded in 'values' represented by each hand. These hands are both weighed and weighted by varied voices, carrying their hope.—S. Eshle

Andrew McLaughlin • Conor Chell
Editors

Building the Case: How Lawyers are Leading on Sustainability

Insights and Inspiration from Leading Lawyers Across North America

Springer

Editors
Andrew McLaughlin
Dieppe, NB, Canada

Conor Chell
Calgary, AB, Canada

ISBN 978-3-031-88454-2 ISBN 978-3-031-88455-9 (eBook)
https://doi.org/10.1007/978-3-031-88455-9

This Springer imprint is published by the registered company Springer Nature Switzerland AG
The registered company address is: Gewerbestrasse 11, 6330 Cham, Switzerland

Foreword

You'd think it would be easy. Sustainability seems an obvious consideration in any environmental, public policy or business decision. By its very nature, the concept involves making sure things last, minimizing costs and protecting society while preserving land, water, and air. It's rational.

But as we've witnessed, it's far from simple. Consider the climate crisis. Decades ago, science identified the problem of human-caused greenhouse gas emissions building up in the atmosphere and causing disruption of weather patterns. This is leading to worsening natural disasters. Climate scientists have offered a checklist of actions to counteract the problem. It's costly, but less so over the long term than doing nothing and letting the problem fester.

Now look at what's involved in getting there: Massive international summits to agree on emission-reduction targets, national policies and regulations for decarbonization, financial incentives to develop the necessary technology, debates on how fast to push the economy into a low-carbon transition, and even pushback from some who question the reality of the problem in the first place. Each one of those is fraught with complications.

Climate is just one part of the sustainability imperative. It encompasses preserving the natural world, ensuring equitable opportunities for everyone in the workplace, promoting human rights and a host of other critical areas. They all need tending to. The question is, how to accomplish them?

The options have become more complex and contentious since the beginning of this decade when the COVID-19 pandemic prodded society into re-examining the range of vulnerabilities we all face. Governments and healthcare agencies scrambled to find ways to keep their populations safe from a virus for which humans had no immunity. It necessitated keeping people apart until the brightest minds in medicine could learn more about the effects and develop vaccines to limit the spread and severity.

That was plenty to deal with, but at the same time, society demanded that leaders in industry and finance use their wherewithal to accelerate the shift to a future with far less CO2 in energy systems. The police killing of George Floyd in the United States galvanized long-standing discontent among Black and Indigenous people and other marginalized groups about unequal—and often illegal—treatment at the hands of law enforcement and business. Companies around the world responded by strengthening diversity, equity and inclusion policies, or establishing new ones.

As it turned out, utopia is elusive. On the environmental front, activists publicized cases of greenwashing—the practice of making false or exaggerated ecological claims—and pointed out cases where corporations trumpeted diversity targets for directors, managers, and employees, but had no concrete plans to achieve them.

Sustainability has also become fodder in the culture wars. Politicians in Republican-controlled U.S. states blamed sustainability initiatives for weakening the fossil-fuel industry and restricting capitalism. Governments in Texas, Florida, West Virginia and elsewhere prohibited institutional investors and banks from using environmental, social, and governance metrics as risk management tools within their jurisdictions. In New Hampshire, some GOP legislators tried, though failed, to make it a crime to knowingly invest state or taxpayer funds using those metrics, punishable by up to 20 years in prison.

That backlash only intensified with the reelection of Donald Trump to the White House. His administration declared war on diversity, equity and inclusion and sought to eradicate numerous environmental policies on an unprecedented scale. Canada, meanwhile, has promoted a wide array of climate-related policies and regulations, from emission-reduction targets to anti-greenwashing legislation. In Europe, regulators are requiring companies to report on double materiality—disclosing details of the environmental risks they face as well as the risks their operations pose to the environment.

This is all to say sustainability is difficult. Navigating the issues requires knowledge and passion among people in science, engineering, finance, politics, and communications. And, at every step, there are legal tangles to be unraveled. If a company wants to start cleanup of a toxic site, lawyers must study all the regulation and property ownership. Perhaps Indigenous communities must be consulted and brought into the project. Who pays? Are affected residents entitled to compensation? Are multiple levels of government involved? Will the project go to a public hearing? Are regulators living up to their obligations? Think of all the riddles to be solved in the names of sustainability and the law.

The legal experts whose essays fill these pages come from all sorts of backgrounds, education, specialties, political stripes and nationalities. Some practice in Canada, others in the United States and Mexico. What they share is keen interest in figuring out how the principles of sustainability fit into the law, and how laws can be improved by promoting positive environmental and societal outcomes.

Some of their names are instantly recognizable from the worlds of politics, academia, and business. Others have worked behind the scenes within large and small law firms and financial institutions, plying their skills in climate-focused finance, regulatory reform and other important areas. In my job as a journalist focused on these issues, I've interviewed some of them and relied on their expertise to help inform my reporting with valuable perspectives on stories often made tricky by legal complexity and shifting norms.

They can help you in a similar way if you are considering taking the plunge into a sustainability-focused corner of the law or expanding your current practice into new challenges. Their essays comprise a spectrum of subjects, such as improving public health care, examining the connections between human rights and business, probing occupational health and safety, and beyond.

The writers tell their stories through their recollections of key cases, offering insights into navigating important principles and how it can all lead to rewarding career achievements.

This is not a legal field that will fade away as long as the world's ills and inequality require attention, with health, happiness and people's ability to thrive in their surroundings at stake. It will never be simple, and that means there will be a steady demand for those who can spot the problems and revel in the prospect of figuring out how to solve them.

Jeffrey Jones is a veteran journalist who has reported on many topics from numerous locales over three decades. As a reporter and columnist with The Globe and Mail, he has covered such business topics as mergers & acquisitions, private equity and energy. In his current role, he has delved into a wide range of sustainability issues, including global climate finance, corporate net-zero and ESG programs, cleantech financing and renewable energy.

The Globe and Mail,

Toronto, ON, Canada
August 29, 2024

Jeffrey Jones

Disclaimer

This book is a collection of chapters written by individual authors, each presenting their own research, analysis, and viewpoints. The views, thoughts, and opinions expressed in this book belong solely to the contributing author of each chapter, and not necessarily to any of the other authors, the author's employer, organization, committee or any other group or individual. The information provided in this text is general in nature and should not be construed as legal advice.

Readers are encouraged to consult with a qualified legal professional for advice specific to their situation. The authors and editors of this book disclaim any liability for any loss or damage caused by reliance on the information contained herein.

Contents

Shining the Spotlight: Lawyers Leading on Sustainability 1
Andrew McLaughlin

Legacy in Practice: A Family-Driven Path to Sustainability Law 9
Conor Chell

The Law, the Planet and the Future 17
Bob Rae

ESG at the Crossroads of Polarization 23
Erin O'Toole

Sustainability Standards and the Practice of Law 41
Michael Torrance

Building a Human Rights Program 51
Jonathan Drimmer

Taming the Uncertainty: The Art & Science of Corporate Human Rights Counsel 67
Yousuf Aftab

Advancing the Power and Water Sectors 85
Aaron Atcheson

Healing Tomorrow: Designing Sustainable Healthcare in a Complicated World 101
Jennifer D. Suess

Indigenous Law and Sustainability 121
Sandy Carpenter

Social Sustainability and Workplace Law 135
Jillian Frank

Beyond the Bar: Purpose, Risk and Innovation 151
David G. Myrol

Dissecting Sustainability: A Critical Approach to the Sustainable Development Discourse 157
Carlos Escoto Carranza

Drop by Drop: Resiliency and the Pursuit of Personal Ideals and Professional Passions 169
Shenandoah Johns

Shining the Spotlight: Lawyers Leading on Sustainability

Andrew McLaughlin

Abstract

In the introductory chapter, co-editor Andrew McLaughlin explains that a principal objective of the book is to inspire and guide aspiring and mid-career lawyers into the realm of sustainability by showcasing real-world examples of lawyers who have made significant impacts in this sphere. By offering insights into the triumphs and setbacks of these legal professionals, the book aims to make seemingly inaccessible career trajectories attainable, thereby encouraging a new generation of impact-focused lawyers. McLaughlin provides a personal narrative, detailing his entry into the sustainability space through lessons learned from a previous career in diplomacy, and in developing and implementing a global Environmental, Social, and Governance (ESG) Framework at his company.

1 Introduction

From the outset, the goal of this book has been to inspire aspiring and mid-career lawyers alike to delve into the sustainability sphere by sharing real-world examples and case studies of leading lawyers who have had an outsized impact in this space. While a young law student or a third-year corporate associate might be deeply fascinated by topics like decarbonization, human rights, or the protection of biodiversity; pursuing a career in these seemingly lofty spheres may at first seem out of reach. This book is intended to bridge the gap by providing a rare insider's glimpse into that reality—warts and all—through spotlighting specific individuals who have left indelible marks during their career. Ideally, through sharing their stories, it will

A. McLaughlin (✉)
Major Drilling Group International Inc., Moncton, NB, Canada
e-mail: andrew.mclaughlin@majordrilling.com

A. McLaughlin, C. Chell (eds.), *Building the Case: How Lawyers are Leading on Sustainability*, https://doi.org/10.1007/978-3-031-88455-9_1

bring these seemingly inaccessible professional trajectories into the realm of the possible to help spur on a bold new generation of impact-focussed lawyers.

In that spirit, I'll share below how I made my foray into this world through parlaying lessons learned from a previous career in diplomacy, and what led me to collaborate with Conor Chell, my co-editor, in launching this book. Finally, I plant a seed for exploring how legal counsel's role is evolving in this increasingly complex and polarized world—a central theme throughout this book.

2 Setting the Stage

Thinking back to right before the pandemic, the term ESG (initially coined in 2004)[1] was entering the mainstream of the corporate world. It was around this time that the company where I work, a publicly traded global drilling services provider in the mining sector, began to turn its sights towards developing an ESG program. While we already had several of the pieces of such a program in play at the time, including a well-established health and safety program and important community engagement initiatives, our challenge was to formalize an ESG framework that would apply across our 15 countries of operation and that was tailored for our particular operational realities.

During a one-on-one lunch with my CEO, we discussed what this might look like and how we needed to make serious headway on ESG. I expressed my strong interest in the topic, hearkening back to my prior career serving as a Canadian diplomat stationed in Mexico. It was during that 3-year assignment that I truly began to appreciate just how important embedding sustainable business practices could be to a company's long-term success.

My role at the Canadian Embassy required me to get out of Mexico City and venture into the states as much as possible. While my primary focus was to report on organized crime and corruption, one of the topics I was drawn to was covering mining interests across the country and the myriad issues at play when a mine is part of a community.

One encounter stands out vividly in my memory from those travels—a meeting with an impressive young Harvard-trained Mexican lawyer who headed up an NGO that tracked alleged abuses by foreign mining companies operating throughout the country. She was cordial, polite, and clearly incredibly sharp. She cited instances where her NGO believed certain foreign companies were falling afoul of their human rights obligations and where weak local governance structures were ill-suited to address these concerns. She and her NGO colleagues had frequently met with Canadian Embassy Officials over the preceding years to share these concerns. By the time we met, I was just the latest in a series of Canadian Embassy representatives she had engaged with. I could tell she felt like she was just going through the

[1] See reference in "The Global Compact, Who Cares Wins: Connecting Financial Markets to a Changing World", https://documents1.worldbank.org/curated/en/280911488968799581/pdf/113237-WP-WhoCaresWins-2004.pdf.

motions in our meeting. And I had very little I could do in my role, except to promise that I would write back a detailed report to Ottawa about our encounter, and that I'd debrief our ambassador.

Despite feeling pretty powerless in that meeting and others like it, my three years in Mexico provided me with a front-row seat to get a deeper understanding of the complex landscape involved when a mining operation is part of a community. These experiences provided a first-hand look at some of the economic and social benefits that such an operation could bring to local communities; but also exposed the environmental and human rights concerns that arise on occasion.

Over the years visiting these communities, it became evident the companies that were poised to be successful were the ones that paid more attention, that listened to these concerns, and that genuinely cared about the communities where they operated. Conversely, those companies that set up with the sole aim to extract as many resources as possible with little regard for anything else, were more likely to run up against strikes, blockades, shutdowns or worse. This essentially boils down to having a social license to operate in a community. While it wasn't an unfamiliar concept for me at the time, it was only upon witnessing what it means on the ground that its significance truly resonated. It's these lessons that have stuck with me to this day, and that continue to inspire my work on sustainability and ESG.

3 The Pandemic Years

Fast-forward back to the lunch with my CEO…

As our discussion drew to a close, it was decided: I would spearhead the development of our company's ESG program. Following a couple of months of internal and external consultations, I presented the draft ESG framework to our Board of Directors. The key components included: a foundational ESG Policy that set out our guiding principles and key priorities (taking into account where we could have more of an impact based on the nature of our operations); the establishment of a Global ESG Committee made up of a diverse cross-section of employees from across our operations and markets; the development of an ESG Repository to track and monitor key metrics; internal and external communications strategies; and ESG oversight at the Board of Directors level. The framework was approved in June 2020—just months after the world was plunged into the throes of the Covid-19 pandemic.

At that time, corporate executives' inboxes across North America and beyond were increasingly inundated with ESG-related commentary and service offerings—a surge that only accelerated with the onset of the pandemic. Concurrently, a corporate marketing frenzy was already well underway with companies across all sectors vying to outdo one another in showcasing their sustainability credentials. Amidst this fervour emerged a looming risk of greenwashing, casting a shadow over genuine sustainability efforts.

One of the most emblematic movements during this time was the global push for companies to make net zero pledges (i.e. corporate commitments to have net zero carbon emissions in their operations by a future date, e.g. 2040 or 2050)—a commendable yet fraught endeavour. Through my discussions on this topic with a number of sustainability professionals, I learned that a significant number of reputable organizations across a breadth of industries had made net zero pledges without a plausible plan for achieving them.

When net zero was discussed at my company, it was clear that making such a pledge would require ignoring our most basic operational realities: our core business is running largely diesel-powered rigs and equipment in some of the most remote regions on earth, oftentimes far from operating mines. Even a battery-powered rig would inevitably need to be frequently recharged by an external power source. For us, while there continues to be a strong commitment to reduce our greenhouse gas (GHG) emissions, banking almost entirely on an unknown future technological breakthrough to bridge the gap to net zero was not a responsible foundation upon which to make such a pledge.

Amidst this landscape of mixed intentions and dubious commitments, a seismic shift in consumer preferences was strengthening nevertheless, marked by a growing appetite for sustainable goods and services (sourced through sustainable supply chains). Yet in this context, there were very few people in the corporate world who were talking openly about the nitty-gritty of implementation—the "how" behind the rhetoric.

I set out to help fill that void.

4 Sharing Our Story

From my perch as the sole lawyer in a small market cap mining services company headquartered in Atlantic Canada, I began writing about our ESG story in various publications—sharing concrete examples of how we were navigating these challenges on the ground. Despite our modest ESG team of two (including myself, and our remarkably talented Sustainability Coordinator, Lisa Holt) and the fact that we were operating within an industry with significant sustainability challenges on the horizon, we raised our hand to signal our commitment to effecting change.

Part of the story involved sharing the steps we'd taken to date, describing our main priorities given our operational realities, highlighting some of the crucial challenges and constraints we were facing, and detailing how we were attempting to tackle these issues. In doing so, I wanted to position our efforts within the broader mining industry picture. While it may not make good fodder for the headlines, there is an important story that goes largely untold about the substantial sustainability work undertaken over the preceding decades in mining. In more recent years, remarkable strides have been made in the industry: from the development of highly sophisticated safety programs and technologies to the proliferation of industry-wide standards aimed at mitigating environmental impacts and protecting human rights. Every day, legions of dedicated personnel across the mining industry around the globe are putting these programs into practice. (As an aside: for impact-focused lawyers, few industries offer such great potential to make a difference, particularly given mining's central role in the green energy transition.)

In sharing our story, I also wanted to take a step back to shed light on some of the more universal issues facing companies across all sectors. For instance, my op-ed in Canada's leading national newspaper drew attention to pressing concerns regarding the ever-expanding ESG Ratings industry.[2] Part of the criticism included pointing out that small sustainability teams like ours were spending much of our limited capacity sifting through the multitude of ratings agency reports to correct errors and misstatements, an exercise that consumed valuable time and energy that could be much better spent on actually improving ESG performance. This piece prompted a number of Chief Legal Officers and Chief Sustainability Officers to reach out to me to share their own perspectives and concerns with the ratings industry.

Another major theme I returned to regularly was legal counsel's critical role in advancing the broader sustainability agenda. For example, in 2020 I argued that in-house counsel are well-positioned to contribute and lead in this area because:[3]

- We enjoy broad visibility and engagement across all of the organization's business lines.
- We are intimately involved in development of corporate policies and standards, as well as in compliance, governance, human rights, anti-corruption, diversity and inclusion, and enterprise risk management (including risk mitigation) matters—all of which are [core ESG topics].
- We are sometimes referred to as the conscience of a corporation, which aligns well with the ESG ethos.
- The General Counsel, as a strategic and trusted business advisor, has a direct line to corporate leadership, which is critical to the effective development, implementation and oversight of [an organization's ESG program and initiatives].

As I shared our ESG and sustainability journey in concrete and practical terms through different publications and speaking engagements, I had a sense that I was striking a chord and getting some unexpected and outsized traction. Through these efforts, I began to engage with a number of lawyers across Canada and beyond who were facing similar challenges, but who had largely tried to address these issues exclusively in-house or in consultation with a small group of external advisors. To make a real difference, it was evident that we had to move beyond these silos to a more constructive, open and collaborative conversation—*an approach that isn't necessarily instinctive in the legal profession*. For instance, I highlighted decarbonization as just one such subject ripe for such discussions:[4]

> …there are certain areas where the in-house bar's default toggle should be set to "open sharing" mode, albeit with the necessary safeguards in place. Decarbonization squarely falls within this category. By collaborating and supporting each other in these efforts, we can help propel our organizations forward while driving broader advancement across industries.

---

[2] McLaughlin (2022).

[3] McLaughlin (2020).

[4] McLaughlin (2023).

What I've since realized is that this call for action on its own without a larger collective push won't move the needle. What's needed is a coming together of the distinct narratives and individual stories to help paint a broader picture of the legal profession's outsized impact in the sustainability landscape—particularly as most of this work is taking place behind the scenes. This is what we're hoping to achieve through this book.

5 Conclusion: Navigating Today's Choppy Waters

When we conceived of this book, the world appeared to be unraveling: war had broken out again in Europe and the Middle East; political polarization and protectionism were becoming further entrenched; year after year of historic floods and wildfires served as stark reminders of the accelerating impacts of climate change; and trust in our institutions was rapidly eroding. The entire post-WWII world order seemed to be barrelling off a cliff.

Yet through another lens, the promise of a better future was approaching on the horizon. There was a sense of being on the cusp of significant changes that had the potential to fundamentally upend the material world as we knew it. Electric vehicles were becoming increasingly commonplace on the roads, albeit still hindered by the inadequacy of charging network infrastructure. The way we heat our homes and our appliances were all undergoing generational if not revolutionary change. The cost of solar energy had plummeted dramatically, signalling a monumental shift in energy production.[5] And through all of these developments, there were important legal constructs, frameworks and principles being challenged and leveraged to advance the agenda—mostly behind the scenes, but at times making headlines.

In this highly volatile moment, many corporations are struggling to navigate these choppy and uncertain waters as we enter a new era and the green energy transition. Given these complexities, more and more corporate leadership are turning to lawyers, both in-house and external counsel, to help explain the changing world around them and to lead the way in navigating the risks and opportunities that lie ahead. Through this, the role of legal counsel is undergoing a dramatic evolution from a primarily technocratic role to that of a skillfully equipped advisor and leader for uncertain times.

In this context, we consider ourselves extremely fortunate to collaborate with many of these leaders as they generously share their stories in the chapters that follow. Through this collective effort, we aim to help inspire a community of practice of impact-focused lawyers across North America and beyond.

[5] The Economist (2024) Sun Machines: Solar, an energy source that gets cheaper and cheaper, is going to be huge. https://www.economist.com/interactive/essay/2024/06/20/solar-power-is-going-to-be-huge.

6 Special Thanks

I'd like to close out with a special note of thanks to my co-editor, Conor Chell, for agreeing to join me in this project and for sharing his compelling family history in his introductory chapter. It's with his inspiring story in mind that we've decided to direct any proceeds from this book to the Salvation Army.

References

McLaughlin A (2020) An ESG Roadmap for In-House Counsel. Canadian Corporate Counsel Association, 4 August 2020. https://www.ccca-accje.org/The-In-House-Edition/Career_List/2020/An-ESG-Roadmap-for-In-House-Counsel

McLaughlin A (2022) Crowded ESG ratings landscape sows confusion for investors. Globe & Mail, 24 April 2022. https://www.theglobeandmail.com/business/commentary/article-crowded-esg-ratings-landscape-sows-confusion-for-investors/

McLaughlin A (2023) In-house counsel's role in decarbonization. Canadian Lawyer Magazine, 14 September 2023. https://www.canadianlawyermag.com/inhouse/news/opinion/in-house-counsels-role-in-decarbonization/379639

Andrew McLaughlin lives in Dieppe, New Brunswick, where he tries to keep up with his two amazing young kids, Pablo and Ana Lucia, alongside his wife, Alix, and their Mexican expat street dog turned suburban princess, Maya. He is the VP–Legal Affairs, General Counsel and Sustainability Lead at Major Drilling Group International Inc. (TSX:MDI), the largest global drilling services provider to the mining industry. He previously served as a diplomat at the Canadian embassies in Havana and Mexico City, and was a Canadian foreign service officer from 2006–2015. Andrew is on the Board of New Brunswick's economic development agency, ONB (a Crown corporation), where he serves as the Vice Chair, and is an Editorial Board member of Canadian Lawyer Magazine. In recent years, Andrew has been making the case for lawyers to lead the charge on sustainability in their organizations, with his publications featured in the Globe and Mail and Canadian Lawyer Magazine (among others) and is a highly sought after presenter on this topic. His op-ed on Canada's critical minerals strategy was featured in the Hill Times and he was cited in the Toronto Star on this topic. His submission on the future of Canadian Diplomacy was highlighted in the Canadian Senate's final report on this topic and was referenced in the National Post. In 2024, he received the Canadian General Counsel Award for mid-market excellence. In 2023, he chaired the ESG Summit that brings together leading sustainability executives from across the country. In 2018, he was named a Lexpert Rising Star for leading Canadian lawyers under 40, and as part of the Banff Forum's "Emerging Canadian Leaders Under 40". Andrew holds a Bachelor of Commerce from Mount Allison University, an LLB from the University of New Brunswick, and an LLM in International Business Law from the University of London (QMUL/UCL).

Legacy in Practice: A Family-Driven Path to Sustainability Law

Conor Chell

Abstract

This chapter explores the deeply personal and professional journey of Conor Chell as he builds the case for sustainability within the legal profession. Starting with a touch of humor, Chell introduces themes of humanity, legacy, and courage. He reflects on his family's tragic and resilient history, highlighting his great-grandfather's courageous stand for miners' safety, the charitable actions of the Salvation Army during a family crisis, and admiration for his parents as educators. These personal anecdotes set the stage for Chell's dedication to sustainability and the ethical practice of law.

Professionally, Chell narrates his transition from an environmental and regulatory lawyer at a prominent national law firm to the first environmental lawyer at Canada's then largest integrated energy producer, detailing the extensive scrutiny and learning he encountered. Inspired by mentors and sustainability's growing global importance, Chell envisioned a new kind of legal practice. Drawing from Malcolm Gladwell's insights on opportunity and cultural legacy, and Seth Godin's principles on market-driven success, Chell established a unique sustainability focused legal practice, first at a large regional law firm, and now as a Partner and National Leader of ESG Law at KPMG Law LLP.

Through these experiences, Chell reveals the importance of integrating personal values with professional aspirations to drive meaningful change in the field of sustainability law and sets the stage to inspire a new generation of lawyers and the vital role they can play in this transformative era.

C. Chell (✉)
KPMG Law LLP, Calgary, AB, Canada
e-mail: conorchell@kpmg.ca

A. McLaughlin, C. Chell (eds.), *Building the Case: How Lawyers are Leading on Sustainability*, https://doi.org/10.1007/978-3-031-88455-9_2

1 Introduction

Any author worth their salt knows that a good story starts with a hook. Since I am not a good writer, but I am a good dad (or try very hard to be), I instead start all of my stories with a dad joke. Here goes: 'Why don't skeletons fight for sustainability?'
Answer: *Because they don't have the guts!!!*

I like this particular joke because, aside from being funny (subjectively speaking perhaps), it touches on three themes that are important to me, both personally and professionally. The first is **humanity**; something that is implicit in sustainability, and something I personally think the profession of law could use more of. The second is **legacy**; what are we leaving behind and how have we made the world and our workplaces, better places for our children and our work families? And the last (but definitely not least) is **courage**. Do we have the courage, as people and as lawyers, to do what is right, and can we act courageously, both in our career choices and in the day-to-day advice we give our clients.

As most of my stories go, this story will be one that blends personal and professional elements, starting with by far the most important of those elements.

2 Sustainability Is Personal

It might sound like an odd place to start, especially after such a fabulous dad joke off the top, but on the personal side, I will begin with my family and going back a few generations, with a family tragedy. On my mother's side, her grandfather (my great grandfather) was a troubled soul and an abusive alcoholic. I won't spend much time on the details other than to say that his life came to an abrupt conclusion on a cold Saskatchewan winter's day, and involved him murdering my great grandmother and then taking his own life. My grandfather, Floyd Chatfield, as the oldest son, was left to pick up the pieces. Given that it was the 1950s in rural Saskatchewan, and given the scandalous nature of the incident, none of the churches in the area were willing to help Floyd bury his parents nor with a funeral. Having no other place to turn, the Salvation Army stepped in and helped my grandfather dig his parents' graves, by hand, and held a funeral ceremony on the barren and frozen prairie. How is that for humanity—helping people who need it the most, when they need it the most? I didn't learn of this story until I was well into my adulthood, as my grandfather Floyd understandably never spoke of it, at least not with me. That said, since learning of this piece of my family history, a significant amount of my charitable donations go to the Salvation Army, and so too shall any proceeds I earn from the co-editing of this book.

Moving along from that uplifting introduction to my family, I will now move on to another indisputable and uniformly accepted fact, which is that the best sustainability lawyers are bred from coal mining genetics. My great grandfather on my father's side was Jim Chell. Jim was the underground mine manager, and later the head mine manager of the Cumberland Mine in Springhill Nova Scotia from 1926 to 1956. Towards the end of his tenure, he wore out his welcome with management as a result of initiating too many 'home runs'. A 'home run' occurred when Jim believed there was an unsafe working condition in the mine, and he would send the workers home, which as you might expect, was a very unpopular decision with management at the time. Referring back to one of the themes I mentioned above, it is hard for me to imagine the courage it took for Jim to make those decisions. Ultimately, his employment was terminated in mid 1956. For those who are familiar with Canadian mining history, you probably know what happened next. Only a few months after Jim's departure, on November 1, 1956, a massive explosion shook the Cumberland Mine, trapping 118 men below surface—88 miners were rescued, while 39 lost their lives. Less than 2 years later, on October 23, 1958, a 'bump' (an underground earthquake) again shook the mine, this time resulting in the loss of 75 lives, which was the largest mining disaster in North American history up until that time. At least nine songs commemorate the event, the most notable being *'The Ballad of Springhill'*, which I remember my Nana Lois singing beautifully and often when I was growing up (and much better than Anne Murray or U2, I might add).

Both of my grandfathers, Floyd Chatfield and Rod Chell (Jim's son), eventually ended up working in mines in northern Manitoba (Flin Flon and Snow Lake) in the mid to late 1950s. They sacrificed their bodies; broke many bones and suffered hearing loss, black lung, strokes and heart attacks, as well as significant mental trauma. If there is any silver lining to be found, it is that my parents met in those small mining towns. My parents, Tom and Jan Chell, were the first on either side of their respective families to go to university, obtain university degrees, and work as professionals; first as teachers, then as principals, then as superintendents of education. I am infinitely proud of them, and even moreso when I hear from people in the communities they worked in, and whose lives they touched in often profound ways. Being educators, humanity was part of their job descriptions, and it took alot of courage to forge the career paths that they did.

Clearly without this lineage, I wouldn't be where I am today and I am deeply thankful for the various sacrifices my family made in order to allow me to live the life I do and pursue the career that I have.

So why am I telling you these things about my family, might you ask? The reason is legacy. As I write this, I am in my mid 40s, a relatively new father of two young boys, aged 18 months and 3 weeks, respectively. Maybe it is a mid-life crisis of sorts; maybe it is a mid-life epiphany, but whatever you want to call it, over the past several years I have found myself increasingly thinking about what my legacy will be. My grandfathers sacrificed their bodies so their children wouldn't have to, and my parents contributed significantly to the communities they served. I often wonder, what sacrifices and contributions have I made? I also wonder what will my children and their children think about me? What stories will they tell? If they tell stories, will they be like those I tell about my parents and grandparents? Will they be proud of me? Will they think I made an impact?

I don't know the answer to all these questions, but I do know that when Andrew McLaughlin approached me to co-edit this book with him, I wasn't going to pass up the opportunity to share part of my story, and my family's story. Hopefully in doing so, I can leave something that forms part of a legacy that my family will be proud of, and perhaps inspire another lawyer or two out there to do the same.

3 Greener Pastures: A Legal Career Rooted in Sustainability

On the professional side, I am an environmental and regulatory lawyer by trade. I started my legal career at the national law firm of Bennett Jones LLP in their Calgary office in 2007. It was a wonderful place to begin a career in law, and I learned a great deal about environmental and regulatory approvals, compliance, and many of the sustainability concepts that have become increasingly more prevalent over the past few decades. I was also afforded the opportunity to work for some of the leading lawyers in Canada on files with some of the most heavily scrutinized companies and industries from a sustainability perspective, such as oil and gas, mining, and agriculture. These early lessons were foundational in shaping me as a lawyer, and in many ways as the person I have become.

In late 2010, I was approached by Suncor Energy to become their first, and at the time, only, environmental and regulatory-specific lawyer. For the next decade plus, I worked on a variety of interesting environmental and occupational health and safety files with many sustainability components, and learned a great deal. Given the industry in which Suncor operates, I was also exposed to some of the most significant scrutiny globally that has been levied on any industry from a sustainability perspective. And most importantly, perhaps, I met an amazing mentor (Shenandoah Johns, now the Chief Sustainability and Environment Officer at West Fraser, and a contributor to this book), who taught me more about integrity and courage than anyone before or since.

Somewhere around 2020, and reflecting on the sustainability legal work I had been exposed to up until that point in my career, I started thinking more and more about my professional legacy, and what I wanted to do with the rest of my life and career. Importantly, I noticed a shift in the collective conscious of people and governments around the world, and the increasing focus on sustainability as a business imperative. I also started to read….a lot. I read books and took courses on sustainability. I read LinkedIn posts from people around the world on sustainability. Sustainability soon became all I thought about, and I began thinking about ways I could turn sustainability into a full time career.

I also read books about life and business, in particular, Malcolm Gladwell's *Outliers*[1] and Seth Godin's *This is Marketing.*[2] In *Outliers*, Gladwell theorizes that opportunity sometimes matters more than hard work or raw talent. Applying that concept to my career, I saw an opportunity, based on where the world was going from a sustainability and legal perspective, to build a unique and successful sustainability practice. Gladwell also discusses cultural legacies, and how cultural

[1] Gladwell (2009).

[2] Godin (2018).

expectations and habits can be a severe hindrance to success. This made me think about several aspects of the traditional culture of law firms and the practice of law, and a few examples of cultural habits that came to mind were the hierarchical structure of law firms, the high costs of service, and the sometimes poor delivery of services (there is certainly a place for a traditional legal memo, but that place isn't as the sole medium by which legal services should be provided, in my humble opinion). With that in mind, and looking to avoid some of the downsides associated with the classic practice of law, I set out to build a different kind of legal practice, one that was sustainability-focused, humanistic and client-centric.

Knowing (or perhaps more accurately at this point, hoping) that a successful sustainability legal practice could exist, I turned my mind to the 'how', and that answer was found by reading Seth Godin's *This is Marketing*. There are many great lessons in *This is Marketing*, but the two that resonated the most with me were: (1) to be successful you must be market-driven, not marketing driven, and (2) to be successful, a better business plan is needed; in particular, a good plan takes a universal need and makes it specific. In the context of sustainability and the law, this meant building a legal practice around incoming mandatory sustainability and ESG legal reporting requirements and anti-greenwashing laws, which at the time were universal needs that Canadian businesses didn't know they needed yet.

Having made my mind up that I would try to create a sustainability legal practice, but having been an in-house lawyer for over a decade at that point, I dusted off my contact list, and got to work. After speaking with several national and regional law firms about the idea of creating a unique ESG/sustainability specific legal practice, I ultimately accepted an offer to create an ESG practice group at MLT Aikins LLP, which is a regional law firm in Western Canada, and a firm I am forever grateful to for giving me the opportunity. Before beginning work with the firm, I conducted thorough market research to inform how I would build my team. I spoke to clients about their needs and expectations and learned that they were looking for comprehensive and holistic solutions to ESG issues that went far beyond addressing purely legal issues alone. Clients wanted a "one-stop shop" for advisory, technical, and legal services. So that's what I built.

But the most important part of the practice is the people. I built our practice based on the feedback received from clients and prospective clients, and I recruited team members who are not only great people, but could also address non-legal issues more efficiently and effectively than lawyers. I first hired a professional civil environmental engineer named Laura Roberts who had significant environmental, regulatory and sustainability experience. The next team member to join was Maya Douglas, who has her Master's in climate change law and has extensive experience as a sustainability and risk advisor and consultant. I am certainly biased and often 'drink my own Kool-Aid', but I believe Laura and Maya are two of the top sustainability advisors in Canada. Most importantly, our team cares deeply about one another and our respective families, friends and loved ones, and I am proud of the humanity we have together infused into our practice.

For over two and a half years at MLT Aikins LLP, Laura, Maya and I worked hard to build our ESG legal practice and brand, and we were very successful in doing so. So much so, that in late 2023, we were approached by KPMG Law LLP with an offer to join the organization and create a National ESG Law team. With the promise of being able to offer an even broader suite of multi-disciplinary ESG & legal services to the top clients in Canada, we joined KPMG in the early part of 2024. Since joining the firm, our practice has taken off and very recently, we also added three fantastic individuals: two sustainability lawyers named Manpreet Dhillon (Vancouver) and Madi Vallani (Toronto), and a senior sustainability consultant named Lauren Forrest (Calgary) to the team, all of whom I have no doubt will be valuable additions and superstars in their own right.

4 The Next Frontier of Sustainability Law

So after providing a taste of my personal and professional journey so far as it relates to sustainability and the practice of law, what comes next? On the personal side, I am looking forward to spending time with my partner Julie and our two sons, Roark and Emery. Having them in my life is the reason my heart beats, and the reason why a career in sustainability law makes so much sense to me. On the professional side of things, myself and my team have much to look forward to in terms of helping companies navigate incoming sustainability and ESG legal risks, and hopefully, guiding companies as they look to improve their ESG and sustainability performance. Most importantly, though, my hope is that we and our clients continue to do work that we can be proud of (and hopefully that our families and friends can be proud of too).

The legal profession is changing rapidly and is ripe for disruption; getting involved in sustainability allows lawyers to participate in that disruption. There are also a wide range of disciplines and approaches that touch on sustainability principles and the law, many of which will be canvassed by the outstanding group of contributors that follow. Speaking of which, thanks very much to my co-editor Andrew McLaughlin—this was your idea after all (and a good one at that—thanks for including me). And to all of our contributors—thank you so much for taking the time to share your stories; they are wonderful and I am in awe of the impact you have each made. Lastly, to all of the aspiring sustainability lawyers out there, I hope this collection of stories provides some inspiration—there is significant opportunity in the sustainability legal space, and on behalf of myself, Andrew, and all of our contributors, we encourage you to seize it.

References

Gladwell M (2009) Outliers. Back Bay Books, New York
Godin S (2018) This is marketing. Portfolio/Penguin, New York

Conor Chell is the father of two boys, Roark and Emery, and along with his partner Julie and their English/Bull Mastiff Mr. Robert Gronkowski, makes his home in the Foothills west of Calgary, Alberta. Conor is one of the top ESG and Sustainability lawyers in Canada with over 20 years of experience. At KPMG Law LLP, Conor is a Partner and the National Leader of ESG Law. Conor leads a collaborative, cross-functional and multi-disciplinary team of sustainability professionals who advise clients on ESG legal risk and disclosure (e.g., greenwashing), responsible supply chain and due diligence (e.g. modern slavery), ESG strategy and program development, and ESG policy development and training. Conor holds a Global Competent Boards Designation (GCB.D) in ESG, a Certificate in Sustainable Finance and ESG from UC Berkeley School of Law and in 2015 was named a Lexpert Rising Star for leading Canadian lawyers under 40. Conor has been twice named a Canadian Environmental Thought Leader by Mondaq (2022 & 2023), speaks frequently across North America on ESG topics, and has been featured in the Globe and Mail, the Financial Post, the Toronto Star, Canadian Lawyer, Investment Executive, DOB Energy, Advisor's Edge, among numerous other publications.

The Law, the Planet and the Future

Bob Rae

Abstract

In this chapter, Bob Rae, Canada's Ambassador and Permanent Representative of Canada to the United Nations in New York, reflects on his ongoing journey through environmental and labour advocacy, highlighting the interconnectedness of sustainable development and social justice. Drawing from his early experiences with labour organizations, Rae underscores the critical need for integrating environmental stewardship with workers' rights. He recounts his political career, emphasizing the creation of the Environmental Bill of Rights and his efforts to promote sustainable economic models. Rae also discusses the global challenges of climate change and biodiversity, advocating for stronger international laws and enforcement mechanisms. Rae concludes with a global call to action that honours both the planet and its people, inspired by indigenous principles of sustainability and intergenerational responsibility.

1 Introduction: Environmental Awakening

I can vividly recall reading Rachel Carson's "Silent Spring" in high school in the 1960s. My high school geography teacher Phil Thomas was the first person to point out to me that the world's glaciers were receding and put us to work assessing what this would mean for water flows in India, Nepal and China. "The planet is not getting healthier" he told us, as we discussed Carson's ideas about the unintended consequences of new chemicals, and how one person's powerful reliance on scientific evidence and courageous dissent could make a difference.

B. Rae (✉)
United Nations, New York, NY, USA

A. McLaughlin, C. Chell (eds.), *Building the Case: How Lawyers are Leading on Sustainability*, https://doi.org/10.1007/978-3-031-88455-9_3

2 Early Influences and Activism

Pollution Probe was the first environmental organization to seize my attention at university, and branches sprung up across the country reminding us that an economy based on externalizing the costs of waste led to the harshest of consequences. The first departments of the environment were created in Canada at the provincial level in the late 1970s, and a deeper political awareness about the consequences of unsustainable growth were widely debated in provincial and federal politics in Canada as I was beginning my political career.

While at law school at this time I worked in a legal aid clinic located in the YMCA building at the corner of College and Dovercourt streets in Toronto. We worked on behalf of a fledgling organization called the Union of Injured Workers, which represented workers injured in industrial plants and construction sites across the city. I also joined the staff of the National Office of the United Steelworkers union part time and taught industrial relations at the University of Toronto. All these experiences made me acutely aware of the link between industrial pollution and workers' health and safety, and how economic and legal theories of "externalities" and reducing the economic burden on employers put workers' lives at risk.

These issues were not just theories or legal arguments, and my perspective on life was changed by what I saw with my own eyes, rather than simply with what I read.

3 Political Career and Environmental Legislation

When I became provincial leader in 1982, the New Democratic Party I had the honour of heading for 13 years took on many critical issues of the day leading to important breakthroughs such as the creation of the Environmental Bill of Rights with an Environmental Ombudsman, and new rules of forest management and conservation that created greater sustainability in an industry that was the economic lifeblood of Northern Ontario. Just before being elected in 1990, I had been arrested on the Red Squirrel Road near the Timiskaming First Nation for joining its Chief, Gary Potts, in a protest against the logging of one of the last, great stands of old growth forest in the province.

That protest was not without its critics, notably BC union leader Jack Munro, and members of his union in Ontario, as well as a lumber industry determined to get maximum value out of timber cutting rights. But I felt that deep concerns about indigenous rights as well as the environmental arguments about how change to a truly sustainable model for the economy could happen, made for an argument we couldn't ignore. We began our work on plans to reduce waste, ensure clean and safe production, and respect the principle that we have to leave the planet in better shape than we found it.

I appointed Maurice Strong, who had been the first director of the UN Environment Program, to be the Chairman and CEO of Ontario Hydro, and he undertook to start a transformation that would put conservation at the heart of energy production in the province, a change that was tragically cut short by the decision of a successor government to cancel all conservation programs, and led to a lost decade in investments in public transit, renewable energy, and progressive approaches to land management and urban planning, deeply damaging initiatives that would set the province back many years.

The government also addressed the health and safety problems that I had seen firsthand in my earlier work with the Union of Injured Workers and the Steelworkers. In addition to dealing with structural and policy problems at the Workers Compensation Board, we also changed health and safety regulations that made it difficult for workers in construction, mining, and industry to prove that long term illnesses were connected to pollution and emissions inside the workplace. Some of the most difficult and compelling cases involved workers suffering from asbestosis and other lung diseases, and the barriers that prevented them from getting the compensation and treatment they needed.

We also changed labour relations laws to change the imbalance between workers and business owners, to make it easier for workers to form and join unions, to require recognition and good faith bargaining in collective bargaining, and to insist on first contracts when agreements were unduly resisted. This is directly connected to improving industrial health and safety, and to saving lives: study after study has shown that unions are key in providing voice and the influence required to get better deals for their workers, not only on wages, but on advocating grievances and improving working conditions.

Being in government between 1990 and 1995 was my first real lesson in the challenges that environmental and labour issues pose to those seeking to put sustainability and health at the heart of political life. Both areas are linked in this important way: employers are prepared to use the "jobs card" when governments press ahead with laws ensuring health and safety protection, bargaining, and emission reductions.

In an economy that was recovering from recession these were powerful arguments, and in my experience employers and corporations were not afraid to use them. But the political challenge goes beyond the arguments and actions of the private sector. I discovered that people are green in general, but less so in particular. The environment was seen as an issue to be dealt with in prosperous times, not during a recession. Farmers resisted regulations on fertilizer and pesticide use, and the resistance to "green taxes" (higher taxes on gas guzzling cars for example) was widespread.

4 Global Perspective on Climate Change

In the eighties and nineties, the policy frame began to include the question of climate change, or as it was first known to me, "global warming". Canada's cold climate made it difficult to campaign in the winter against something called "warming", but that was not then, and is not now, the greatest political conundrum. Global climate change is a global process whose impacts are now unquestionable, but while immediate problems are increasingly clear (flooding, forest fires, dramatic shifts in climate patterns), there is still a sense in the land that a global problem will require more dramatic changes from large political actors like the United States, China, India, and other rapidly industrializing countries, and that as a small emitter Canada really can't do much.

Mark Carney has pointed that in addition to the well-known "tragedy of the commons" there is now a "tragedy of the horizon"—the focus on targets that need to be met in 25 years means that current voters (and parties pandering to these instincts) are apt to say that the easiest thing to do is to keep pushing the costs down to future generations. This has its parallel in Canada's—and the provinces'—dramatic failure to tax the current use of resources properly, and put the funds raised in a savings bank for future generations to use to diversify the economy in years to come. Norway has been the one country in the world that stands out for its environmental and fiscal prudence. Its National Wealth Fund is now worth over a trillion dollars. Like Aesop's fable of the two squirrels—one who squandered and one who saved—there are far more squanderers than savers in the world.

It was the great eighteenth century conservative thinker, Edmund Burke, who challenged the simple "social contract" argument of liberal and radical thought that government was essentially a contract between the government and the governed. He insisted that this was too limited a view, and insisted that rather government was about the contract between generations, including those have died and those who are yet to be born.

When I began working again in northern Ontario for the Mattawa Tribal Council, nine first nations struggling to establish a new relationship with the Crown, I began hearing from the Anishinaabe Elders about the principle of "seven generations thinking". Long before Burke thought of his formula, there was a realization deeply imbedded in indigenous thinking that all of us who are alive are keepers of a flame, stewards of lands, air, and water whose central task is to ensure these common goods are not irreparably damaged or destroyed, but are passed on to children and grandchildren in better shape than we found it, and hence the concept of sustainability, ensuring the circular strengthening of economies, rather than an embrace of private accumulation in the hands of a few, is firmly embedded in Canada's own indigenous history and identity.

There is a profound difference between seeking a new relationship between humans and the planet and the transactional thinking and politics that still dominates much Western thinking about politics, economics, and society. In my work with Indigenous communities I became convinced that the cord of indigenous governance that had been brutally cut by the arrogance of settler power had to be restored, because it has so much to teach us about our relationship to the land, to nature, and to life itself.

Just as my work in the politics of Ontario was cut short by a new government with a new set of values and policies that I felt ran deeply against the long term interests of the province, my work as a negotiator with the Mattawa Tribal Council was ended by a new Conservative in Ontario which felt that opening the region to business development with a new set of transactional deals was the right way to go. Cutting short environmental assessments, dividing Indigenous communities and working closely with mining companies to ensure speedy development became the order of the day.

Shortly after this happened, I was asked to become Canada's Ambassador to the United Nations in New York. This has opened a new chapter and has brought together a number of strands that have woven their way through my life. Coming to New York in the middle of the Covid crisis was a shocking experience in itself. A new virus spread quickly from a market (or a lab, the controversy continues…) in China to become a global (that word again) pandemic, and by the spring of 2020 had led to a series of decisions that shut down the world's economy.

The world has known slumps and depressions before, and has also known pandemics before. The severe flu which hit the world in 1918 ended up killing more people than died in the conflagration of World War I. But as dramatic as those events were, and as devastating as previous epidemics have been, the remedy applied to the global economy was dramatic.

As it so happened, when I came to New York in August of 2020 I was immediately thrust into the work of the United Nations, which had kept on meeting virtually, with little interruption, through the pandemic. The day after I started work, I was on my first zoom call, and the zoom calls, and now many in person meetings, have continued. One of my first areas of work was to co-chair, with the Jamaican Ambassador, meetings on a project on Financing for Development, in which the anger and frustration about the unequal impact of climate change was taking us deeper into a deep divide between haves and have-nots, with the risk that the clamour would prevent the world from reaching a common will to action.

The level of common awareness on climate change and the growing threat to biodiversity, but the challenge before us all is that this awareness is not leading to change to match the full dimensions of the difficulties the planet is facing. Our global talk is far greater than our global walk. Setting targets is one thing. Meeting targets is something else. The decisions of governments, companies, and people to meet and exceed these targets just aren't happening at a pace to bend the curve. The age-old dilemma is still with us: laws without enforcement are powerless to do the job, and enforcement only happens when there is the political will to back it up.

The "political will" has now to become global. National targets are insufficient to do what needs to be done. We have to create the global institutions that will get the world to where it needs to be, and where it wants to be. This is no easy job. It will not happen in a single leap. Innovative technological change, courage and determination at the local level, and a pragmatic willingness to make change happen: all are required, and more will need to be done.

5 International Law and Future Directions

In particular, my experiences have led me to a deeper appreciation of the challenge facing international law more broadly, as well as the ongoing necessity of turning our ideas into reality. My own journey toward law school, and away from more academic pursuits, was driven by my sense that there was a healthy granularity to the law that forced me to add a stronger dose of pragmatism to my approach to social and economic issues.

I have never regretted this decision, and the law has provided me with a vital framework of how to move from theory to practice. Many environmental issues are, of necessity, issues that transcend political boundaries and jurisdictions. The French philosopher Pascal reminded us all that justice without enforcement is powerless, but enforcement without justice is another name for tyranny. Nowhere is this axiom more compelling than in the global discussions about climate change and biodiversity.

The Treaties and Conventions that form the ever-growing web that is international law are only as strong as their ability to be enforced. This is certainly true about climate change. Countries agree to targets. But there is no international body that has the power to enforce targets. As the reality of the climate crisis sinks in, it will be increasingly important to move beyond the world of consensus and voluntary compliance to the world of norms and laws that are legally enforceable, and that are met with measures equal to the task.

The good news for lawyers is that the future will be full of opportunities to respond to these growing crises. We must put an end to impunity, or the health of present and future generations is at risk. Pascal's conundrum needs to be resolved in favour of justice and enforcement if we are to fulfill our obligations to humanity and the planet itself.

Hon. Bob Rae is the Ambassador and Permanent Representative of Canada to the United Nations in New York. Since assuming this role, he has been active in all aspects of the work of the UN, culminating in his election to the Presidency of the Economic and Social Council (ECOSOC) for 2024–2025. He served as Premier of Ontario from 1990–1995, interim Leader of the Liberal Party of Canada from 2011–2013 and was appointed as Canada's Special Envoy to Myanmar (2017) and Canada's Special Envoy on Humanitarian and Refugee Issues (2020). From 1996 to 2008, he was a partner at Goodmans LLP and wrote government reports on the Air India bombing and higher education in Ontario. He also served as Chair of the Toronto Symphony Orchestra, the Royal Conservatory of Music, the Institute for Research in Public Policy and the Forum of Federations. Mr. Rae taught law and public policy at the University of Toronto and in the VicOne programme at Victoria College, and was a partner and senior counsel to the law firm OKT LLP from 2014 to 2020, specializing in indigenous law and constitutional issues. Bob Rae is a Privy Councillor, a Companion of the Order of Canada, and a member of the Order of Ontario.

ESG at the Crossroads of Polarization

Erin O'Toole

Abstract

This chapter explores the evolving landscape of Environmental, Social, and Governance (ESG) initiatives amidst rising global polarization. Hon. Erin O'Toole, President & Managing Director of ADIT North America and former leader of the Conservative Party of Canada, examines how geopolitical shifts, economic changes, and cultural forces are challenging the legitimacy and effectiveness of ESG. The chapter traces the historical development of ESG from corporate governance and Corporate Social Responsibility (CSR) initiatives, highlighting key milestones such as the Dey Report in Canada and the UN Global Compact. It delves into the impact of deglobalization, some of the causes of anti-trade populism, and social media on ESG, illustrating how these factors contribute to the current polarization.

O'Toole argues that for ESG to remain relevant and effective, it must adapt to contemporary geopolitical realities and engage a broader range of stakeholders, including the working class and non-elite voices. The chapter concludes with practical recommendations for evolving ESG practices, emphasizing the need for inclusive dialogue, realistic energy transition policies, and the integration of Indigenous reconciliation into ESG frameworks in Canada.

1 Introduction

Environmental, Social, and Governance (ESG) is facing an identity crisis thanks to immense, global geopolitical changes. To put it bluntly, ESG's legitimacy is under attack as the world is being reshaped by a variety of economic, cultural and political

E. O'Toole (✉)
ADIT North America, Toronto, ON, Canada
e-mail: erin.otoole@adit-na.com

A. McLaughlin, C. Chell (eds.), *Building the Case: How Lawyers are Leading on Sustainability*, https://doi.org/10.1007/978-3-031-88455-9_4

forces. The positive impacts of ESG can only be maintained if proponents acknowledge this and adapt the dialogue to meet the times. If not, ESG initiatives will continue to be negatively impacted by geopolitics and polarization, eroding progress on sustainability initiatives. This has already begun, and if it continues, ESG may be looked back upon one day as just another acronym that fell victim to the culture wars of the 2020s.

ESG has always been private sector led, growing from the natural evolution of corporate governance and corporate social responsibility (CSR) initiatives. Corporate leaders in the west wanted to find the right balance between the free market pursuit of profit and the public good for the workers and the communities the business operated in. This is how ESG developed as both a responsible reporting tool and an investment philosophy.

ESG must now evolve as a tool to better reflect the world today, and not the world as it was or how we would like it to be. Globalization is unwinding as political populism blames trade and corporate elites for social decline. Some environmental and labour policies can be gamed by our foreign rivals and turn western ESG altruism into a weakness. And excessive ESG ambition can become an impediment to wealth creation and a further contributor to political polarization. To ensure that it survives, ESG needs to reform itself to meet the challenges of the times. How business leaders and their advisors rise to meet the challenges facing ESG is the goal of this chapter.

2 Canada and the Corporate Governance Revolution

In the 1990s and early 2000s, corporate governance in democratic capitalist countries, including Canada, underwent a major transformation. Executives and boards of directors faced increasing scrutiny and their advisors—lawyers, accountants, and consultants—pushed for change.

In Canada, this process was led by the Ontario Securities Commission (OSC) under the chairmanship of Peter Dey, a prominent corporate lawyer. In response to governance failures, the OSC released the groundbreaking 1994 report *Where Were the Directors?*,[1] commonly known as the Dey Report.

The *Dey Report* emphasized that corporate boards must not only oversee business strategy but also act as stewards of the company by upholding ethical standards and managing external risks. This report marked a pivotal moment in Canadian corporate governance, introducing the idea that directors should take into account factors beyond profitability, such as the ethical and social impacts of corporate decisions. The Dey Report also laid the groundwork for the development of Corporate Social Responsibility (CSR), which would later evolve into ESG.

For the first time in Canada, a major report on corporate leadership stated that the "moral and ethical tone" of the business was a critical component of how the business was managed.

[1] https://chartthefuture.ca/assets/uploads/img/Where-Were-the-Directors-the-Dey-Report-optimized.pdf.

The *Dey Report* became the roadmap. By the late 1990s, many corporations had implemented governance reforms that reflected the growing focus on professional and independent oversight of management. These changes, largely driven by private sector demand for better governance and risk management, also signaled an increasing awareness of external factors such as environmental and social concerns that could impact business. This approach would become central to the evolution of ESG, as companies realized the importance of addressing not just financial risks but broader societal issues as well.

3 Global Governance and Sustainability

Around the same time that the *Dey Report* was transforming corporate stewardship in Canada, discussions on sustainability were evolving. In 2000, the United Nations (UN) held the Millennium Summit to set ambitious goals for human rights, labour standards, transparency, and environmental improvements, leading to the creation of the Millennium Development Goals (MDGs). The summit also launched the UN Global Compact, a public-private partnership aimed at integrating sustainability into the private sector, incorporating many of the governance and CSR reforms from the 1990s into its work, emphasizing market incentives, social entrepreneurship, and sustainable investing.

By 2015, when the MDGs concluded, the momentum for sustainability was so strong that the initiative became permanent through the Sustainable Development Goals (SDGs). The SDGs set the standard for global stewardship and inspired projects like the Global Reporting Initiative (GRI) and the Carbon Disclosure Project (CDP). These initiatives put pressure on corporations to integrate sustainability and broader social reporting into their operations. Management teams and boards adopted more sophisticated practices to track and mitigate external risks.

Despite political attacks on sustainability efforts in recent years, SDGs were driven by the private sector, not by some 'globalist' agenda as some conspiracists would suggest. These goals emerged naturally from the evolution of corporate governance and the generational shift in leadership within democratic capitalism. However, one major challenge remained: while corporate boardrooms embraced sustainability, the broader working class and everyday employees were largely left out of the conversation.

4 ESG Is Born[2]

The term "ESG" first appeared in the 2004 report *Who Cares Wins*, published by the UN Global Compact. The report emphasized that long-term investment success depends on the stability of a sustainable economy and society. It argued that

[2] See https://documents1.worldbank.org/curated/en/280911488968799581/pdf/113237-WP-WhoCaresWins-2004.pdf.

integrating environmental, social, and governance factors into financial decision-making would create more stable markets and reduce risk for investors.

Who Cares Wins was the creation of many finance firms, such as Goldman Sachs and Morgan Stanley, who were already factoring in environmental and social issues in their investment strategies. The report reframed these factors as not only responsible business practices but also as profit-driven strategies. This marked the beginning of ESG's transformation from a corporate social responsibility (CSR) initiative into a broader investment philosophy aimed at creating value for investors alongside more sustainable and predictable capital markets. There was always a profit motive at the centre of the development of ESG. It was geopolitical change, however, that led to the ESG identity crisis.

5 Deglobalization

The rise of ESG paralleled other political and cultural forces in the west, placing it in conflict with growing political polarization. As CSR initiatives transformed into ESG in the late 1990s and early 2000s, attitudes toward trade and globalization began to shift in the U.S. and other western democracies. Over the last two decades, the post-World War II consensus on free trade has been challenged by great power rivalries. Non-market economies like China began to challenge democratic market economies like the United States and the result has been a global political realignment. Globalization is gradually unwinding and is being replaced by managed trade between countries.

Trade has become increasingly regulated through tariffs, non-tariff barriers, and security sanctions aimed at protecting key sectors and countering geopolitical rivals. The last decade has seen tariffs and sanctions imposed on critical commodities like steel, aluminum, critical minerals, and semiconductors, essential to the decarbonization of the global economy. The world is now multi-polar, with countries vying to control the decarbonization agenda and challenge American dominance.

The shift to managed trade has led to the creation of modern trade blocks that reinforce geopolitical tensions. The G7, led by the U.S., has traditionally promoted trade, aid, and security collaboration, while its rival, the BRIC countries (Brazil, Russia, India, China and others),[3] has evolved into a serious political and economic rival.[4] China's Belt and Road Initiative has extended its influence in the developing world, leaving the G7 scrambling to catch up. As trade and geopolitical alliances continue to evolve, this realignment will shape global norms for years to come.

Deglobalization has even affected close trading partnerships like Canada and the U.S. Once vaunted trade deals like NAFTA have been renegotiated because of geopolitical pressures. The Americans have also pushed Canada to align on managed trade initiatives to address perceived risks from China. The Obama administration pressured both the Harper and Trudeau governments to ban Huawei equipment from

[3] https://www.goldmansachs.com/our-firm/history/moments/2001-brics.

[4] https://www.visualcapitalist.com/charted-comparing-the-gdp-of-brics-and-the-g7-countries/.

Canada's 5G telecommunications infrastructure and expressed frustration over Chinese takeovers of Canadian defence companies.[5]

The Trump administration dramatically upended global trade, beginning with its withdrawal from the Trans-Pacific Partnership (TPP) in 2017. Shortly after, President Trump announced the renegotiation of NAFTA, followed by new tariffs on steel and aluminum imports, and justified it as necessary for U.S. national security.[6] While Canada and Mexico were first excluded from the application of these tariffs, it was clear that this reprieve was temporary.

Canada had long been warned that it was being viewed as a 'backdoor' for Chinese exports to the U.S. As the Conservative Shadow Minister for Foreign Affairs, I heard this message consistently in discussions with my American counterparts. These frustrations led many American lawmakers to believe that Canada should face the same tariffs imposed on other countries.

My meetings with both Democrats and Republicans revealed a growing, bipartisan frustration. with Canada due to our lack of seriousness and general unwillingness to align on evolving U.S. trade and security priorities. When Canada was granted a temporary exemption from the application of tariffs, I warned that this should not be seen as a victory. Canada was still losing the wider battle for permanent exemption, as the U.S. increasingly expected alignment on trade and security matters.

> We have to be a deep and full partner, and when the prime minister goes down and talks about his progressive agenda or caribou herds – all of those things are important – but they take a secondary priority to market access, to security. They [the Trudeau government] have to realize the administration they're dealing with has priorities and if Canada doesn't line our priorities up with theirs, we're going to pay a price.[7]

Two weeks later, the Trudeau government introduced new trade measures that were intended to address the transshipment of steel and aluminum through Canada. Despite these last-minute efforts, the Trump administration ended up proceeding with steel and aluminum tariffs against Canada. Instead of clearly committing to align with U.S. trade policies regarding China, the Trudeau government doubled down and responded with retaliatory tariffs against American exports. At the time, Prime Minister Trudeau said he wanted to stand up to the U.S. and said that "common sense will prevail".[8]

This period of time also featured the renegotiation of NAFTA, and the Trudeau government's "progressive trade agenda" that was aimed to contrast with Trump's regressive image by focusing on social issues rather than finding common ground

[5] https://www.theglobeandmail.com/news/politics/liberals-reverse-course-on-chinese-deal/article34441975/.

[6] https://www.govinfo.gov/content/pkg/STATUTE-76/pdf/STATUTE-76-Pg872.pdf.

[7] https://www.cbc.ca/news/politics/trump-tariff-steel-1.4567730.

[8] https://www.pm.gc.ca/en/news/speeches/2018/05/31/remarks-prime-minister-canada-steel-and-aluminum-tariffs-imposed-united.

on critical sectors of the economy like auto manufacturing.[9] The result of this strategy was devastating for Canada and led to our exclusion from final negotiations between the U.S. and Mexico. A bilateral deal was reached and Canada was forced to capitulate to terms already decided.[10] The new NAFTA—now dubbed the Canada U.S. Mexico Agreement (CUSMA)—almost missed having the 'C' because of Canadian posturing and our unwillingness to align on trade and security concerns.

While Trump lost the 2020 presidential election, the Biden administration continued the same approach with respect to China and managed trade. Within weeks of taking office, Biden issued an executive order to safeguard American supply chains,[11] leading to increased investment in critical minerals[12] and tariffs on Chinese manufactured electric vehicles.[13]

In 2020, as Conservative Party leader, I began to talk about the need for Canada to align with the U.S. on trade and security with respect to China. We needed to recognize that globalization and the 'rules-based order' were changing, especially in areas like critical commodities, energy, minerals and defence technology. During an interview with the *Globe & Mail,* I argued that Canada should prioritize trade with democratic countries that respect ESG standards and be cautious with our approach to non-market economies.

> We have to recognize that people have been left behind, and globalization has been used by some countries that don't follow the rules-based system that we've talked about. So, I would like to see free trade amongst free countries that follow the rules. And I do think this type of alignment is coming. It also means conservatives can't just be free-traders regardless of whether China's operating internment camps for part of their population. Our values should never be for sale. If you're a democratic country that cares about the environment and social governance and Indigenous engagement, you should only use Canadian energy, softwood and rare-earth minerals. That's how we should be marketing ourselves.[14]

In August 2024, after many years of avoiding these issues, the Trudeau government decided to mirrored the American approach and finally applied tariffs on steel, aluminum and electric vehicles.[15] This decision ended up being made in the final

[9] https://www.canada.ca/en/global-affairs/news/2017/08/address_by_foreignaffairsministeronthe-modernizationofthenorthame.html.

[10] https://www.nytimes.com/2018/08/27/us/politics/us-mexico-nafta-deal.html.

[11] https://www.whitehouse.gov/briefing-room/presidential-actions/2021/02/24/executive-order-on-americas-supply-chains/.

[12] https://www.whitehouse.gov/briefing-room/statements-releases/2024/09/20/fact-sheet-biden-harris-administration-takes-further-action-to-strengthen-and-secure-critical-mineral-supply-chains/.

[13] https://www.whitehouse.gov/briefing-room/statements-releases/2024/05/14/fact-sheet-president-biden-takes-action-to-protect-american-workers-and-businesses-from-chinas-unfair-trade-practices/.

[14] https://www.theglobeandmail.com/politics/article-jobs-at-core-of-otooles-vision-for-canadas-next-decade/.

[15] https://www.canada.ca/en/department-finance/news/2024/08/surtax-on-imports-of-steel-and-aluminum-products-from-china.html.

months of the Biden administration, and just a few months before the re-election of Donald Trump. The Trudeau government was hoping that 'better late than never' could help improve bi-lateral relations, but that has certainly not been the case.

6 Globalization Leads to Polarization

To understand the origins of the realignment of trade we are experiencing you must go back to the 1992 presidential election in the United States when third party candidate Ross Perot warned about the dangers of trade deals like NAFTA, predicting a "giant sucking sound" of jobs moving to Mexico. Perot lost the election but was the most successful third-party candidate in modern U.S. history largely because of his strong concerns about trade and job losses on the American economy.

As Western society grappled with the consequences of trade liberalization, large segments of the population realized they hadn't benefited equally. Manufacturing jobs were shifted to lower cost jurisdictions, leaving many American workers vulnerable to job loss or chronic underemployment. This trend accelerated when the U.S. supported the accession of China into the World Trade Organization (WTO) in 2001. What NAFTA had started, China's WTO entry intensified, without sufficient policy shifts to retrain people and redevelop the affected communities.

The U.S. trade deficit ballooned during the 1990s and 2000s, with low-cost imports flooding in from countries like China and Mexico.[16] These jurisdictions had lower environmental and labour standards, making their goods cheaper to produce than in the U.S. or Canada. To compound matters, wages for blue collar and non-managerial workers remained largely flat for decades. Although political rhetoric often touted a shift toward a knowledge economy, programs to support displaced workers never materialized, leaving real wages for many not far removed from 1970s levels.[17]

This economic decline paralleled a major loss of public trust in government and civic institutions, particularly in the so-called 'rust belt' states of the Midwest and Great Lakes regions. Politically important states like Ohio, Michigan and Pennsylvania, once manufacturing powerhouses, saw their industries decline, and with them, their political landscapes shifted. Once Democratic strongholds due to their industrial histories and unionized workforces, these states became politically volatile as economic hardships deepened.

The term of 'rust belt' itself came from the politics of declining fortunes. In the 1984 presidential campaign, Democratic candidate Walter Mondale compared the rusting out areas of the Great Lakes and Midwest to the depression-era Great Plains states that had suffered through the famine and hardship of the dust bowl.[18] The journalists on the Mondale campaign turned his "rust bowl" comments into the "rust belt" geographic description we know today, which is how this pejorative economic

[16] https://www.visualcapitalist.com/cp/charted-u-s-trade-deficit/.

[17] https://www.weforum.org/stories/2019/04/50-years-of-us-wages-in-one-chart/.

[18] https://www.britannica.com/place/Rust-Belt.

term has come to define an entire region of the country, symbolizing the fading American dream for millions of residents.

The decline of traditional blue-collar industries like steel production and automobile manufacturing eroded the social contract that once promised economic stability and upward mobility for working-class Americans. Jobs that had been passed down through generations were no longer available, forcing many young people to leave for opportunities elsewhere. In 1940s America, over 90% of children would do better economically than their parents, but in the last 20 years that number has plummeted to around 50%.[19]

As high-paying, stable jobs disappeared, workers faced lower wages, contract work, and diminished benefits. Unionization rates fell and many were forced into low-wage, unskilled jobs in the service industry. According to the U.S. Census Bureau, the rate of Americans holding more than one job has increased steadily over the last 20 years with women holding multiple jobs at a higher rate than men.[20]

Alongside declining job prospects, social challenges began to steadily trend higher, with rising disability claims, drug addiction, and crime plaguing these struggling communities. Rural America, in particular, saw disability rates skyrocket, with one in three residents in some areas identifying as disabled. The opioid epidemic further devastated these communities, turning a social issue into a national health crisis.[21]

To borrow language from the American culture wars, globalization and 'the rules-based international order' only seemed to be helping the 'coastal elites' and wealthy Americans. The 'rules' seemed to be stacked against millions of working-class Americans. The disconnect between political leaders and millions of Americans who were struggling began to erode the political consensus on the issue of trade and foreshadow the populism that is upending politics.

During the 2008 U.S. presidential campaign, then-candidate Barack Obama acknowledged the resentment felt by rust belt voters, noting that their bitterness stemmed from decades of unfulfilled promises. In his infamous "*cling to guns or religion*" comment, Obama recognized that both parties had failed these Americans when manufacturing was shifted overseas without a plan for those who were left behind. But his condescending tone alienated many, reinforcing the feeling that politicians were out of touch with their struggles. Despite apologizing, Obama doubled down on his message, reinforcing the idea that both parties had failed to address the impact of globalization on working-class Americans. This acknowledgment was a precursor to the populist movements that would soon dominate U.S. politics.

> But I will never walk away from the larger point that I was trying to make. For the last several decades, people in small towns and cities and rural areas all across this country have

[19] https://opportunityinsights.org/national_trends/.

[20] https://www.census.gov/library/stories/2021/02/new-way-to-measure-how-many-americans-work-more-than-one-job.html.

[21] https://www.cdc.gov/ncbddd/disabilityandhealth/features/disability-prevalence-rural-urban.html#:~:text=According%20to%20this%20study%2C%20the,compared%20to%20large%20metropolitan%20areas.

> seen globalization change the rules of the game on them. When I began my career as an organizer on the South Side of Chicago, I saw what happens when the local steel mill shuts its doors and moves overseas. You don't just lose the jobs in the mill, you start losing jobs and businesses throughout the community. The streets are emptier. The schools suffer.[22]

American political leadership was beginning to acknowledge the deep-seated resentment growing in large parts of the U.S. with respect to globalization and the loss of jobs and financial security for millions of working-class Americans.

7 The Great Recession Deepens Polarization

The subprime mortgage crisis and Great Recession from 2007 to 2009 marked the beginning of Obama's presidency and made his campaign promises more challenging to fulfill. The economic damage was severe for lower- and middle-class families, with billions in housing wealth wiped out and many losing their homes. The government's bailout of banks in 2008 and large automakers in 2009 helped stave off even worse economic outcomes, but the policy choices also deepened class divides. While corporations bailed out by taxpayers saw record profits and raising executive salaries a few years later, millions of Americans who were poorer because of the housing crisis saw their wages barely increase. Within these fertile soils of resentment grew the start of the severe political polarization that we see today.

From 2009 to 2013, the Tea Party emerged as a populist political movement, transforming U.S. politics. It created a strong anti-establishment wing within the Republican Party, challenging traditional positions like support for free trade, trust in institutions and environmental protections. The Tea Party began to influence primary contests and promoted policies that were far removed from some of the more traditional Republican positions going back to President Reagan. These shifts reflected the frustrations of many Americans who felt left behind by both major parties.

While the Obama administration took steps to address inequality and health care reform, it did not prevent growing bitterness from turning into populist anger. Working-class voters in rural America and the rust belt were worried about rising drug use, crime, and the lack of job prospects for their children. The Great Recession had them increasingly worried about losing their homes and their future. While these voters saw their lives and communities in deep decline, the party that was supposed to be fighting for them—the Democratic Party—was shifting into a period of intense focus on global issues and social policies such as climate change, international aid, immigration, Diversity, Equity and Inclusion (DEI) initiatives and controversial issues like defunding the police. These priorities felt disconnected from the struggles many working-class Americans were facing.

[22] http://obamaspeeches.com/E06-Barack-Obama-AP-Annual-Luncheon-Washington-DC-April-14-2008-religion-guns-pennsylvania.htm.

By the end of the Obama's presidency, millions believed they were left out of political discussions and left out of the prosperity of America. This discontent created fertile ground for a politician that would appear to feel their pain and 'blow up' the system that they believed was working against them.

When Donald Trump descended down the golden escalator in Trump Tower on June 16, 2015, to announce his run for President, the decades of political frustrations and rising polarization crystalized, and politics has never been the same. Within seconds, Trump framed his candidacy around trade, particularly with China, and painted a picture of America in decline.

> Our country is in serious trouble. We don't have victories anymore. We used to have victories, but we don't have them. When was the last time anybody saw us beating, let's say, China in a trade deal? They kill us. I beat China all the time. All the time.[23]

Trump's campaign was built on decades of rising frustrations and solidified a new political movement, the Make America Great Again (MAGA) wing of the Republican party, which propelled him through the Republican primaries and into the White House in 2016. While Trump lost the 2020 election, the clear victory in the electoral college and popular vote in 2024 showed that the MAGA wing of the Republican party is now the entire party.

8 Polarization in Canada and Europe

In Canada, the economic and cultural trends seen in the U.S. were mirrored, albeit with less intensity and on a delayed timeline. The decline of manufacturing in the Great Lakes region impacted communities like Windsor, Hamilton, Sault Ste. Marie and Oshawa (which I represented in Parliament). Although Canada weathered the Great Recession better than the U.S., owing to high energy prices and housing stability, similar populist pressures began to emerge by the mid-2010s.

Justin Trudeau's 2015 election victory, based on messages of 'Hope and Hard Work' closely echoed themes from the Obama presidency at a moment when Canada seemed immune from populist pressures. This would soon change with the election of Donald Trump in 2016 and challenges the Canadian economy would soon undergo that would begin to cast clouds on Prime Minister Trudeau's "sunny ways."

As a Conservative opposition critic and candidate for the party leadership during this period, I began to see real signs of polarization in Canada. Rising populist anger was particularly visible among energy sector workers in Western Canada and blue-collar voters in rural parts of Ontario. By the end of 2018, Canada had lived through the cancellation of several pipeline projects, the imposition of tariffs and the loss of investment in large projects because of changes to environmental assessment reviews. Global oil prices were bottoming out, and the Trudeau government seemed

[23] https://youtu.be/aV-nyOM4_gc.

to be over emphasizing climate change policy, which led many in the west to believe the challenges they were facing were completely the fault of the Trudeau government. Empty office towers in Calgary came to symbolize the decline of the mighty Alberta economy and blame—fair or not—was laid entirely at the feet of Ottawa.

In my own community, the end of 2018 also saw the announcement by General Motors that the auto assembly plant in Oshawa was closing after serving as the heart of the auto sector in Ontario for generations. My father had worked at GM Oshawa, so like thousands of families we were shocked by this news. Much like in the U.S., many Canadians were beginning to feel disconnected from the priorities of the government and felt left behind or forgotten by politicians.

Europe also experienced rising populism, driven by economic insecurity and a perceived disconnect between politicians and the needs of ordinary citizens. The Brexit movement in 2016 almost mirrored the rise of the Tea Party in the U.S., with traditional 'red wall' Labour voters in industrial towns aligning with Conservatives to vote for leaving the European Union. Boris Johnson's election as Prime Minister on the back of shifting political allegiances was a direct result of this political realignment.

By 2018, rising populist tensions in France took the form of the '*Gilets Jaune*' (Yellow Vest) protests. These protests, sparked by rising fuel costs and environmental taxes, represented the frustrations of rural and working-class French citizens who felt disconnected from urban elites. Similar trends were seen in Germany and other parts of Europe, where far right parties were able to tap a growing frustration about the disconnect between the urban policy makers and the needs of many rural families and communities.

By late 2018, Canada was seeing its own 'Yellow Vest' protests in western Canada, culminating in the "United We Roll" truck convoy to Parliament Hill in February 2019.[24] Many participants were workers in the energy sector, hit hard by low oil prices and government policies. Some of these individuals also became involved in western separatist movements, expressing frustrations over economic challenges and a growing disconnect with Ottawa.

When the COVID-19 pandemic hit in 2020, the same organizers were able to rally thousands of new supporters by questioning the need for lockdowns and vaccine mandates. After more than a year of isolation and pandemic restrictions, many Canadians sought a return to normalcy, which gave rise to the "Freedom Convoy" trucker protest in early 2022. While the protest was meant to push back against vaccine mandates, it quickly became a symbol of broader dissatisfaction with the government and fatigue with COVID measures.[25]

The Freedom Convoy was an inflection point for polarization in Canada. Its anti-authority narrative and illegal blockades, including blocking border crossings and bridges, led to heightened tensions across the country. While the government's invocation of the *Emergency Measures Act* to address the protest was an excessive

[24] https://www.cbc.ca/radio/frontburner/what-does-united-we-roll-stand-for-1.5025431.

[25] https://edmonton.ctvnews.ca/united-we-roll-counter-protest-cancelled-after-rail-blockade-cleared-out-1.4819372?cache=havfyladuxaywoyv.

overreaction to this law breaking and anti-social behaviour, the incident underscored the growing divisions in Canada. Polarization had become a defining feature of Canadian politics.

9 Polarization Fueled by Mistrust & Social Media

The one constant between elements of the MAGA movement, Brexit, the Yellow Vests and the Freedom Convoy is a deep and growing mistrust of authority. This grew from the perception that the policy elites in government and business were out of touch from the real needs of citizens, who were being held back economically and socially. There was also a deep sense that the elites looked down on people with different points of view on policies ranging from carbon taxes to vaccine mandates. This same mistrust extends to elites in business and academia and explains why anti-corporate and anti-ESG sentiments are on the rise.

The rise in this angry, anti-authoritarian populism was also happening in a time when the growing influence of social media channels amplifies frustrations and allows protest movements and groups to grow more rapidly. Social media channels allow people to find others who share their views on a range of issues in ways they could never do before. Social media also has the negative impact of constantly hardening and reinforcing people's views because the algorithms have been designed to continue to serve up outrage until the user's view on an issue becomes baked in. Over time, some susceptible users can be molded by the algorithms to move from disagreeing with people with opposing viewpoints to eventually disliking to even hating people with opposing views. Rage posts fuelled by the algorithms have been accelerating the polarization of politics. A study of these issues on Facebook confirmed that "[t]he most abusive people on Facebook, it turns out, are given the most power to shape what Facebook is".[26] This explains how previously off-limits provocations like calling someone a 'traitor' are now almost milquetoast, and how "Fuck Trudeau" comments on Facebook were transformed into a cottage industry of outrage flags, shirts and bumper stickers. I was astounded by encountering people who would have these signs or flags displayed on their homes with young children being raised in that environment. In the worst cases, polarization has normalized the breakdown of civic order and civility.

Politicians and business leaders must understand this acceleration of polarization on social media and avoid instances of dividing political or policy questions into us versus them showdowns. The "cling to guns and religion" comment from Obama in 2008 came before social media could make it go viral, but the "basket of deplorables" comment from presidential candidate Hillary Clinton in 2016 did go viral and became a badge of honour for some of those voters.[27] We saw the same thing happened in France when Prime Minister Édouard Philippe called yellow vest

[26] https://www.theatlantic.com/technology/archive/2022/02/facebook-hate-speech-misinformation-superusers/621617/.

[27] https://time.com/4486502/hillary-clinton-basket-of-deplorables-transcript/.

protestors "thugs"[28], and in Canada when Prime Minister Trudeau called anti-lockdown protestors "racists" and "misogynists" in the middle of the 2021 Canadian election.[29]

10 Polarization of ESG

It should be no surprise that ESG has been drawn into the culture wars given its intersection with issues like climate change, social policy and internationalism that are issues that have been deeply polarized over the last decade. Whether in the Governor's Mansion in Florida or the tractor-filled streets of Paris, the politicization of ESG and its constituent elements has fundamentally changed the rules of the game and companies must realize this.

Polarization means that the 'E' (Environmental) and 'S' (Social) objectives of ESG are being politicized by both left and right. The polarized right believes ESG is nothing more than 'woke capitalism' or even "*communism rebranded*"[30] according to business titan Elon Musk. To the polarized left, ESG is rapidly becoming viewed as nothing more than a form of corporate 'greenwashing' to avoid real action on the environment. In the words of celebrated youth climate activist Greta Thunberg, the last global climate meetings were nothing more than an opportunity for "*greenwashing, lying and cheating*"[31] for those in attendance. With one side claiming there is a climate apocalypse and the other suggesting climate change policies are part of a 'globalist' conspiracy to force us to eat bugs,[32] it is no wonder that normal people are keeping their heads down and trying not to attract controversy on ESG initiatives.

Polarization is even starting to force some corporations to backtrack on some of their existing ESG commitments. In 2023, one of the world's most prominent investors and advocates for the development of ESG as an investment philosophy banned the use of the ESG acronym at his US$9.1 trillion firm. Larry Fink, the CEO of Blackrock, took this extraordinary action after several Republican-led states began to consider investment boycotts of Blackrock funds because of Blackrock's commitment to ESG.[33] He is not alone. The high-profile Glasgow Financial Alliance for Net Zero (GFANZ) banking initiative spearheaded by Canadian Prime Minister

[28] https://foreignpolicy.com/2019/11/29/emmanuel-macrons-france-yellow-jackets-police-europe year of cracking-heads/.

[29] https://www.theglobeandmail.com/politics/article-liberal-leader-justin-trudeau-pledges-legal-protection-for-businesses/.

[30] https://finance.yahoo.com/news/public-being-lied-elon-musk-110000615.html.

[31] https://www.reuters.com/business/environment/greta-thunberg-cop27-an-opportunity-greenwashing-lying-cheating-2022-10-30/.

[32] https://www.npr.org/2023/03/31/1167550482/how-a-conspiracy-theory-about-eating-bugs-made-its-way-to-international-politics.

[33] https://www.reuters.com/business/environment/blackrocks-fink-says-hes-stopped-using-weaponised-term-esg-2023-06-26/.

Mark Carney has been slowly falling apart over the last 2 years as banks backtrack or withdraw because of declining confidence in the initiative and unrealistic climate targets. One departing German banking executive even declared the GFANZ initiative to be nothing more than "Net Zero Blah Blah of ESG LaLa Land".[34] In 2024, European consumer goods giant and early ESG pioneer Unilever backtracked publicly from its ESG commitments. The CEO cited the changing geopolitical landscape as a reason for the change.

> When you have a huge drought for a number of months but everything else is going fine, the attention is on climate. These days it's about wars and rightly so, that's at the forefront.[35]

In Canada, ESG has not yet reached the full culture war stage, but that may not be far off if an economic recession hits and polarization continues. Many of the large Canadian banks have withdrawn from GFANZ and have toned down their ESG language and reporting. A recent survey of Canadian institutional investors managing C$5.4 trillion in assets confirmed that the industry is becoming more cautious in their approach to ESG. Almost a quarter of the firms surveyed confirmed they are in the process of reviewing how they communicate surrounding ESG.[36]

11 Adapt or Die: Where Does ESG Go from Here?

Rather than ignoring the political and economic realities of the era we are in, dialogue around ESG and the public good must adapt to meet the times. This means ESG advocates in academia, the corporate world, law and politics must try and take the issues that have disrupted the ESG conversation seriously and learn from them. Geopolitical realities when it comes to trade, security and diplomacy, and the state of polarized politics in western democracies cannot be ignored. Most importantly, we must ensure that discussions about ESG and DEI actively engage voices outside of elite corridors and the corporate mainstream.

With this in mind, I have a few suggestions that I hope will help inform the necessary evolution of ESG in this age of disruption and realignment.

12 'E': Getting the Energy Transition Right

Corporate Canada and most ESG policy makers are over their skis when it comes to the energy transition and public support. Large segments of the population are not engaged in the conversation on climate change policy and some Canadians even see

[34] https://www.capitalmonitor.ai/analysis/gfanz-why-gls-banks-departure-is-a-serious-setback/?cf-view.

[35] https://www.theguardian.com/business/2024/apr/19/unilever-to-scale-back-environmental-and-social-pledges.

[36] https://www.advisor.ca/investments/market-insights/esg-investment-isnt-dead-priorities-are-changing-millani/.

phrases like ‘climate action’ as a threat to their way of life. The impact of national policies related to environmental assessment and carbon emissions will be inordinately difficult for Alberta and Saskatchewan and this is already straining national unity.

First, Canadian investment and pension funds should not divest from investments in Canadian energy. This is particularly the case for public pension funds who have both a fiduciary duty to generate strong returns for pensioners and a national profile related to the public good that requires public confidence. ESG plans have been too aggressive in the last decade and have been based largely on corporate altruism and unachievable objectives. This has led to lower returns for investment funds and has contributed to polarization and an erosion of public trust in the pension sector. It is time to rebalance the conversation on the energy transition.

Second, the departure of Prime Minister Trudeau from federal politics also led to jettisoning of the federal carbon tax. Despite his longtime support for carbon pricing, Prime Minister Carney saw how divisive the carbon tax had become and ‘axed’ it before the Conservatives could. This is further confirmation that the public has not been adequately engaged in debates about the climate and the economy and was not supportive of public policy that the government believed there was widespread support for. Corporate leaders must adapt to this reality and ensure that industry groups and their ESG plans work to collaborate with a new national approach to the reduction of carbon emissions.

Third, Canadian industry should promote dialogue on the use of Carbon Border Adjustment Tariffs, which is a managed trade tool that the Americans are currently exploring.[37] This would ensure that Canadian steel, aluminum and other carbon intensive commodities or products are not placed at a disadvantage to competitive imports from jurisdictions with higher carbon emissions and hazy environmental disclosures to go along with their lower costs. Canada is one of the world’s largest aluminum producers and our carbon intensity is one of the lowest due to hydroelectricity and investments by our producers.[38] We should work with the U.S. to ensure that we align on these sectors of the economy and not allow bad actor countries to gain advantage due to our pursuit of ESG goals.

13 ‘S’: Social Progress Our Workers Can See

The largest driver of political polarization in the west has been the perceived unfairness of trade deals and the loss of manufacturing jobs to jurisdictions with vastly lower wage rates and lower environmental, labour and social standards. There are a few ways that corporate ESG programs and wider policy discussions can drive

[37] https://www.whitehouse.gov/briefing-room/speeches-remarks/2024/04/16/remarks-as-prepared-for-john-podesta-columbia-global-energy-summit/.

[38] https://natural-resources.canada.ca/our-natural-resources/minerals-mining/mining-data-statistics-and-analysis/minerals-metals-facts/aluminum-facts/20510.

social progress and allow working-class voters to see their views reflected in public policy.

First, consider having a union or employee representative on the ESG committee or board of directors of the company. This is quite standard practice for large companies in Germany, which has had great success keeping their workforce productive and economy competitive.[39] Engagement of workers will also go a long way towards demystifying ESG for non-executive employees, which is a key element needed in reducing the levels of polarization we see today.

Second, ensure that your corporate ESG policy has a commitment to strong labour practices in the company alongside strong and transparent reporting mechanisms for your global supply chain. In 2021, the U.S. Congress passed the *Uyghur Forced Labor Prevention Act* to purge U.S. supply chains of slave labour from China. This legislation has already stopped shipments of goods worth more than US$3 Billion for review or trade enforcement.[40] Canada launched its own version of this legislation with the *Fighting Against Forced Labour and Child Labour in Supply Chains Act*, which came into force in January 2024. A few years prior to this, the Trudeau government opened the Office of the Canadian Ombudsperson for Responsible Enterprise, which expanded the investigatory powers of the Office of the CSR Counsellor for the Extractives Sector that the Conservative government had created. Most Canadian corporations know little about these agencies or the reporting obligations that are now required by law. Corporate leaders must ensure that ESG policies and reporting practices reflect strong support for a supply chain that is transparent and free from human rights abuses or poor labour or environmental practices overseas.

14 'G': Reforming Governance at Home and Internationally

Polarization is going to be the reality for the next few years. Ignoring the concerns or skepticism of large swathes of the population is what helped fuel this phenomenon.

Corporate leaders should push for dramatic reform at institutions like the UN Global Compact and the World Economic Forum. Western countries that value ESG, DEI and other progressive corporate initiatives should ensure that bad actor countries do not take advantage of these values in democratic, market economies. Corporations should push this with their governments. Corporate ESG policies should be more accessible and relatable to employees and the wider public, and should strive to ensure that they do not make polarization worse.

[39] https://worker-participation.eu/legislation/european-company-se/countries-transposition/germany#:~:text=Employee%20representation%20at%20board%20level,companies%20with%20more%20than%202%2C000.

[40] https://www.csis.org/events/new-frontiers-uflpa-enforcement-fireside-chat-dhs-secretary-alejandro-mayorkas.

15 'I': Canadianize ESG by Adding Indigenous Reconciliation

Finally, I have long advocated that ESG policies in Canada should be ESG-I policies to incorporate the concept of Indigenous reconciliation as part of the broader social policy goals of ESG. This is particularly the case when policies touch on development or the extraction or use of energy, minerals, water or other resources. By adding a distinct economic reconciliation focus to ESG in Canada, I believe there will be more opportunity for Indigenous peoples in our country, and more meaningful social license to help get big projects done in this country. From critical mineral mines to electricity generation and transmission to pipeline construction, ESG-I can help make progress on the journey of reconciliation and build a stronger, more self-sufficient and a more equitable country for all Canadians.

Hon. Erin O'Toole is the President and Managing Director of ADIT North America and is the former Leader of the Conservative Party of Canada.

Erin is one of Canada's most respected thought leaders on geopolitical and economic policy issues from his decade of service as a Member of the Canadian Parliament, Minister, Parliamentary Secretary and Leader of the Official Opposition. Prior to his election to Parliament, Erin served in the Canadian Armed Forces and worked for a decade in the private sector as a corporate lawyer and advisor.

In government, he served as the Parliamentary Secretary for International Trade during final negotiations on Canada's historic free trade agreement with the European Union. He was involved in trade negotiations with South Korea, Honduras, and countries within the Trans-Pacific Partnership. Erin also led efforts related to the extractives sector and regulatory cooperation initiatives between Canada and the United States. He also served as Minister of Veterans Affairs and improved service standards and mental health support for Veterans and their families. In opposition, Erin served as the Shadow Minister of Foreign Affairs before becoming the Leader of the Conservative Party of Canada and Leader of the Official Opposition from 2020 to 2022. In 2021, Erin ran for Prime Minister of Canada in a general election held amid the COVID-19 pandemic. He won the popular vote and advocated for a series of economic and geopolitical policies that continue to resonate in Canadian policy debates today.

A graduate of the Royal Military College of Canada, Erin was commissioned as an officer in the Royal Canadian Air Force and flew as a navigator on the Sea King helicopter, which deployed with the Royal Canadian Navy. He was awarded the Canadian Forces Decoration for his service and received a Sikorsky Helicopter Rescue Award.

Following his service in the military, Erin graduated from Dalhousie Law School and worked as a corporate lawyer and business advisor at two of Canada's largest law firms and as in house counsel to one of the world's largest consumer goods companies.

Erin and his wife Rebecca have two children (Mollie and Jack) and a lazy yellow labrador retriever named Wexford.

Sustainability Standards and the Practice of Law

Michael Torrance

Abstract

The connection between sustainability and the practice of law has never been clear—but it can no longer be denied that sustainability presents both an opportunity and a challenge for business leaders and their legal advisors that warrants the attention of the legal profession. Sustainability arises most prominently in areas of corporate governance that are not well defined in traditional legal analysis. For example, the role of the corporation to manage its own adverse external impacts on the environment and society, where minimum legal compliance is insufficient to meet the reasonable expectations of critical stakeholders. Some take this characteristic to mean that sustainability is unrelated to law. The author does not believe this is the case and has built a career around integrating an understanding of sustainability practices as part of broader legal expectations or strategic management. In this chapter the author discusses his journey as a sustainability lawyer and shares some thoughts on the evolving relationship between sustainability and law, in theory and in practice.

1 Introduction

The connection between sustainability and the practice of law has never been clear—but it can't be denied that sustainability presents both an opportunity and a challenge for business leaders, that warrants the attention of the legal profession. Sustainability arises most prominently in areas of corporate governance that are not well defined in traditional legal analysis, where minimum legal compliance is insufficient to meet the reasonable expectations of critical stakeholders. Some take this

M. Torrance (✉)
BMO Financial Group, Toronto, ON, Canada
e-mail: michael.torrance@bmo.com

A. McLaughlin, C. Chell (eds.), *Building the Case: How Lawyers are Leading on Sustainability*, https://doi.org/10.1007/978-3-031-88455-9_5

characteristic of a "beyond compliance" approach to sustainability to mean that the phenomenon is unrelated to law. I don't believe this is the case and have built a career around integrating an understanding of sustainability practices as part of advice on legal expectations and strategic management. In this chapter I will discuss my journey as a sustainability lawyer and share some thoughts on the evolving relationship between sustainability and law, in theory and in practice.

2 My Sustainability Journey

"I want to start a new practice group focused on Corporate Social Responsibility," I said to our Managing Partner as a third year Employment & Labour law Associate in 2008. I argued this was an emerging area where we could advise clients on how to meet international standards in some of the same topics we advised on in domestic law—environmental management, labour standards, health and safety, human rights. "As long as you get your real work done, I'm okay with you working on this." There it began, a tepid endorsement of a vision that has since defined my career.

With the support of my firm, I pursued a Master of Laws (LLM) researching the relationship between law and sustainability standards, publishing a thesis on the topic of the relationship between law and international sustainability standards.[1] I then organized a group of my colleagues who were topical experts on environmental law, human rights, Indigenous rights, health and safety, to write a book[2] on one of the leading international sustainability standards: the IFC Performance Standards on Environmental and Social Sustainability (the "IFC Performance Standards"). The IFC Performance Standards were being used by the World Bank in its own financing activity, and by private banks in project finance agreements around the world, referenced as a standard to be met as a condition of financing. I later did a one-year secondment with an Asia Pacific (APAC) focused Occupational Health and Safety team based in Sydney Australia, where I had the opportunity to work on thought leadership and client briefings for APAC project finance and banking lawyers, making some great contacts and learning a lot about the financial sector and its intersection with sustainability through frameworks like the Equator Principles. Upon my return to Canada in 2013, I made progress monetizing the business of sustainability focused advisory within a law firm, continuing thought leadership initiatives on sustainability and finance, focusing on emerging disclosure related regulation in Europe and elsewhere, advising on the risks related to sustainability claims (i.e. greenwash), and the emerging human rights framework known as the UN Guiding Principles on Business and Human Rights. By 2015 the firm had successfully obtained substantial mandates advising on supply chain human rights due diligence and responding to shareholder proposals calling for adherence to sustainability standards. We had refined a path to building a legal advisory business on sustainability. But the pace of growth was slow and the opportunities limited, with

[1] Torrance (2010).

[2] Torrance (2012, 2021).

legal practitioners playing a limited role in advising on sustainability topics at the time.

Knowing however that I wanted to focus my career on sustainability, and less on litigation, I kept my eyes open for other opportunities. In 2017 I was told about a Director of Environmental, Social and Governance (ESG) role opening up at Bank of Montreal (BMO). The role reported into the General Counsel of the bank and they were looking for a lawyer with a sustainability background. Having just made partner after a long 12-year slog, I was hesitant. But I decided it would give me a chance to manage sustainability from the inside, rather than having to win one-off advisory mandates and never seeing the "big picture" as external counsel. I coincidentally met the then BMO General Counsel, Simon Fish, who I felt had a great vision for sustainability as a legal topic and I was sold on the idea of moving in-house. I convinced the team that I was the right candidate for the role, and I made the transition from private practice to company counsel. It was the best decision I ever made from a career perspective.

I started in January 2018. A few months before, the Financial Stability Board had published the Task Force on Climate Related Financial Disclosure (TCFD) report and recommendations. From that point forward the pace of change for the financial sector on the topic of sustainability was intense. Whereas sustainability had been an afterthought in the industry, a "nice-to-have" leading to reporting that was aligned to voluntary frameworks like the Global Reporting Initiative (GRI), within five years the topic was top of mind for the Executive Committee and Board of Directors. It was also increasingly a focus of regulators and ESG minded investors and other core stakeholders. As the bank's sustainability lead I helped organize and launch a sustainable finance strategy, issue BMO's first sustainability bond, develop an environmental and social risk corporate policy and related management program, developed Enterprise wide training, pioneer a climate scenario analysis program, successfully respond to multiple investor proposals and win numerous sustainability awards, including being listed on the Dow Jones Sustainability World Index for the first time in the Bank's history. In 2021 we launched the BMO Climate Ambition, a net zero aligned strategy for the bank cutting across risk management, commercial opportunity and sustainability commitments. As part of that effort we launched the BMO Climate Institute, a vehicle for thought leadership and engagement on sustainability, and incorporated climate into our core business strategy for the Enterprise, eventually developing and delivering climate training to tens of thousands of bank employees including bespoke training for our Commercial Banking teams. I became Chief Sustainability Officer (the first person to hold that role at a bank in Canada) and my team multiplied in size. With the acquisition of a carbon credit business in 2021 we grew to have dozens of sustainability or climate focused professionals working across the bank. This was a pace of change on a singular topic rarely, if ever, seen in the 200-year history of the bank.

As of writing, the pace of change continues to be brisk. The International Financial Reporting Standards (IFRS) Foundation has published its International Sustainability Standards Board (ISSB) guidelines on climate and sustainability related financial disclosure which is likely to transform financial reporting globally.

Regulators in Europe, Canada and the United States, including prudential banking regulators, continue to focus on the management of climate risk and sustainability. Prudential banking regulators are requiring sustainability and climate disclosure and the adoption and implementation of the TCFD model of climate management, involving explicit approaches to strategy, governance, risk management and metrics and targets. So-called "anti-ESG" backlash has created inverse legal risks for companies where the perception that companies are doing too much on sustainability, or taking measures that are not tied to business considerations, can create legal or political risks. New regulation has also been introduced in more "pro" ESG jurisdictions, to crack down on sustainability claims amounting to greenwash. This includes within competition law in Canada, with criminal sanctions or the prospect of significant penalties. Investors remain interested in the topic with multiple shareholder proposals on ESG topics every year, some receiving substantial investor support.

With these developments and more, the legal context of sustainability becomes clearer every day. The complexity of the field is growing and the legal and business risks of doing too much or too little are becoming more apparent, necessitating a strong corporate governance approach. But still, the connection is often not well understood, and the legal profession is not seen as offering a coherent or market leading (compared to accounting or management consultancies) advisory offering on the topic. This can stem from a lack of understanding of how to advise in the sustainability sphere, and the connection between sustainability and law.

3 'Sustainability Has Nothing to Do With Law': Confronting the Myth

While in hindsight the journey from law firm Associate to Chief Sustainability Officer makes sense, it didn't look that way in 2008 when I started my sustainability journey. The most often heard comment when I discussed sustainability as a legal topic was "sustainability has nothing to do with law". This expressed a view that sustainability was voluntary, not mandatory like law, and therefore completely distinct from law or legal considerations. Sustainability was viewed as a measure of a corporate performance, not a minimal compliance expectation. It was a technical engineering topic, requiring technical engineering skills to understand, not knowledge of rules or the use of advocacy skills. Moreover, since sustainability expectations didn't come from courts or legislatures, or were based on "values" (and therefore derived from subjective value judgments), it could not be related to legal expectations. For each of these reasons, and more, lawyers and non-lawyers alike agreed, sustainability was a "law free zone", unrelated and entirely distinct phenomena. As such, there was no conceivable role for the legal profession to play in the field of sustainability.[3]

I could not then, and do not now, accept these premises which both underestimate the highly consequential nature of sustainability standards of conduct and also

[3] See Torrance (2011).

ignore the relationship between legal systems and international sustainability standards, which become the equivalent of codified industry good practice or custom that play an important role in legal analysis.[4] Based on both my research and experience, I have concluded that the right focus for lawyers is (1) to be expert in the substance of sustainability standards that define the corporate governance of sustainable companies; (2) understand the legal "hooks" for sustainability standards, including where standards are or may be incorporated into regulatory expectation, contracts or form part of due diligence processes; (3) be advisors on the strategic side of sustainability, particularly where it involves the risk of stakeholder conflict and reputation management and corporate governance.

4 How Lawyers Can Be Leading Sustainability Advisors

The clearest connection to the practice of law is through the legal "hooks" for sustainability standards, which are many and varied. Where such hooks exist, it is particularly incumbent on legal counsel to understand the practice of sustainability, which informs the application of legal expectation. International sustainability standards drive sustainability management practices. These include disclosure standards like the GRI and now ISSB, management system standards including those of the ISO, sustainable finance standards of the International Capital Markets Association (ICMA), or more substantive codes like the IFC Performance Standards or the TCFD. These standards and the practices they describe are the core "knowledge" of the sustainability practitioner. There is every reason for strategic legal advisors to be well versed in these standards, to help their clients navigate regulatory expectations, promote strong corporate governance, and manage potential stakeholder conflict and reputational matters. Ignoring these standards in the practice of law does a disservice to the client.

In recent years we have seen international sustainability standards directly inform or even be incorporated into regulation. Examples include the EU Green Deal for Europe, with regulation including the Non-Financial Reporting Directive (NFRD), the Corporate Sustainability Reporting Directive (CSRD), the Directive on Corporate Sustainability Due Diligence (CSDD) and other regulation defining the scope and definition of sustainable finance, like the EU Sustainable Finance Taxonomy. These regulations are based on, and in some cases directly incorporate, international sustainability standards that have been used for a number of years predating these regulations. With the integration of IFRS and other enhanced reporting standards into corporate practices, securities lawyers outside of Europe will need to advise on concepts of materiality taking such standards into account.

In Canada, the Competition Act was updated in 2024 to impose overt prohibitions and the prospect of penalties on companies that make sustainability claims that cannot be substantiated. In some cases, the legislation provides a "due diligence" defence to greenwash, where a company can establish that its claims are based on

[4] See Torrance (2011).

adherence to a relevant international sustainability standard. Adoption of international sustainability standards into legal systems through these mechanisms can involve government playing the role of "meta-regulator" of what would otherwise be self-regulation or standard setting processes that define the parameter of good practice in sustainability.

Aside from direct legislative incorporation or endorsement, utilization within judicial and regulatory processes is another mechanism by which sustainability standards may come to be incorporated into legal systems. Regulatory offences providing for "strict liability," for example in the health and safety or environmental context, may also require corporations to establish "due diligence" if they are to avoid sanction once the factual elements of the offence have been established. Alternatively, tort claims may necessitate the demonstration of "reasonable care" in relation to persons affected by corporate conduct. Due diligence in this context requires that a company show that its approach conformed to internationally accepted good practice and management practices.[5]

Under many occupational health and safety legal regimes, including in the United States and Canada, laws set an expectation that employers will carry out "due diligence" in taking reasonable precautions to protect the health and safety of workers. Failure to meet the due diligence expectation can cause liability within such regimes. In some occupational health and safety statutes, due diligence requires employers to take all reasonable precautions in the circumstances to prevent injuries or accidents in the workplace. This can also mean taking precautions that are not only possible but that are suitable or rational, given the situation. Generally, to carry out due diligence, an employer must implement a plan to identify possible workplace hazards and carry out the appropriate corrective action to prevent accidents or injuries arising from those hazards. The method and process for due diligence, often linked to adoption of international sustainability standards, is therefore essential in understanding the legal standard to be met.[6]

France's Duty of Vigilance Law provides another example of the incorporation of sustainability related due diligence into legal expectations. Adopted in 2017, companies subject to the law must now establish mechanisms to prevent human rights violations and environmental impacts throughout their chain of production, including those of their subsidiaries and companies under their control. These mechanisms must be reported each year as part of a "vigilance plan." UK, Canadian and US modern slavery legislation is not as prescriptive, instead requiring disclosure of management approaches without requiring the adoption of any specific approach to management. However the requirement to disclose creates a "compliance" pull for companies to adopt international sustainability standards into their corporate governance and to communicate to key stakeholders and the market that these issues are adequately managed. Regardless of whether due diligence is legally mandated, any due diligence process involves the identification of facts and the evaluation of those facts in light of an applicable standard, policy or expectation. A

[5] Drimmer and Torrance (2020).

[6] Drimmer and Torrance (2020).

due diligence process is not mechanical and requires the exercise of informed judgment by a competent person. There is an important role for legal counsel in this process.

Beyond specific legislative incorporation, international sustainability standards that are viewed as "best practices" within a certain industry may influence the availability of permits or licenses for corporations or affect regulatory approvals or sanctions. This can be seen in the role that adherence to concepts like "Free Prior and Informed Consent" have played in obtaining approval for major projects in Canada, even where that international standard was not expressly a requirement of "hard law".

International sustainability standards may also become incorporated into the terms and conditions of contractual relationships, as they are in project finance agreements subject to the Equator Principles framework for environmental and social risk management. A corporation may agree to adhere to international sustainability standards in agreements with regulators, commercial suppliers, or as part of membership in an association. Such voluntary undertakings may be converted into a legally binding requirement when a party to such an agreement seeks judicial enforcement of the agreement. Any lawyer involved in the creation of such contracts should have the competence to advise on the standards being incorporated into their clients' contract, though in practice many don't.

Corporate directors may even be held to open-textured legal standards of care, skill and loyalty that can require effective implementation of sustainability standards. The topic of climate change provides an example, where standards like the TCFD are being used by investors and regulators as a good practice benchmark that should be implemented with oversight of corporate boards. This role for international sustainability standards arises where such norms provide a generally accepted set of standards to guide decision making, departure from which is taken as a sign of weak governance. The core elements of recommended climate-related financial disclosures under the TCFD include: (1) governance, which includes the organization's governance around climate-related risks and opportunities; (2) strategy, which considers the actual and potential impacts of climate-related risks and opportunities on business strategy and financial planning; (3) risk management, which includes the processes used by the organization to identify, assess and manage climate-related risks; and (4) metrics and targets, used to assess and manage relevant climate-related risks and opportunities.

Even when it comes to so-called "anti-ESG" legislation in some US states and emerging at the Federal level, the interpretation and application of such laws should not involve the wholesale elimination of sustainability practices or "green hushing". There is a great need to understand what the regulatory concern is, specifically, for example that corporate practices are not based on purely politicized value judgments. The counterpoint to this values-based approach to sustainability is understanding the business value created by adoption of international sustainability standards. There often is a strong business case that fully justifies sustainable business practices. In fact, the business case will often be so strong that abandoning sustainability practices to appease an anti-ESG regulator will often create significant risks to the corporation, its management and board. Legal counsel without sufficient grounding in sustainability strategy or the substance of international sustainability standards will struggle to give strategically valuable advice in interpreting and applying such regulatory requirements.

What, practically speaking, should lawyers do to be better advisors? As a strategic advisor, the lawyer should put themselves in the role of management or the board. It is not particularly valuable to get an "on the one hand, on the other hand" analysis when confronted with hard challenges, tough decisions, or significant reputation risks that go beyond legal compliance. To play that strategic role, the legal counsel must understand not only the strict legal requirement, but the underlying corporate strategy that should drive business decision making. This is all part of the legal standard of "sound business judgment" that directors and officers must strive to meet. To do this in the context of sustainability, a lawyer must be aware not only of hard law but the underlying sustainability standards and good practices of corporate governance that define how a well-managed company should function. Without that connection, legal advice becomes academic, or even worse, harmful to the interests of the company and its leadership.

5 Concluding Thoughts

For legal practitioners to play a strategic role in advising their corporate clients, it is critical to have in-depth expertise in both the underlying sustainability standards and the mechanisms for legal integration. This requires an understanding of the underlying international sustainability standards that must be implemented to meet both stakeholder and legal expectations, expertise on the legal "hooks" for sustainability and connection of sustainability to strategy and corporate governance advice. With this approach, legal advisors can and will be trusted counsel for the senior leadership and directors of companies, large and small. My own career has focused on mastering sustainability standards as a core aspect of my own subject-matter expertise, understanding the integration of such standards into legal standards of conduct, and positioning advice as part of overarching corporate governance and strategic objectives of the company. I would encourage any young lawyer seeking to focus on sustainability to take a similar approach. More seasoned practitioners should also see sustainability as an opportunity to increase their value add as a strategic advisor to clients.

References

Drimmer J, Torrance M (2020) CSR risk management and due diligence. In: Corporate social responsibility, sustainable business. Wolters Kluwer, The Netherlands, pp 603–620

Torrance M (2010) Persuasive authority beyond the state: a theoretical analysis of transnational corporate social responsibility norms as legal reasons within positive legal systems (October 4, 2010). https://ssrn.com/abstract=1687282 or https://doi.org/10.2139/ssrn.1687282

Torrance M (2011) Persuasive authority beyond the state: a theoretical analysis of transnational corporate social responsibility norms as legal reasons within positive legal systems. German Law J 12(8):1573–1636

Torrance M (2012) IFC performance standards on environmental & social sustainability: a guidebook. LexisNexis, Markham

Torrance M (2021) Equator principles and performance standards on environmental & social sustainability, 2nd edn. LexisNexis, Markham

Michael Torrance is Chief Sustainability Officer at BMO Financial Group. He leads sustainability strategy and oversees the BMO Climate Institute, which seeks to bridge the complexities of climate science and policy with business strategy and economics. Prior to joining BMO Michael was a Partner at a major international law firm where he founded the firm's environmental and social risk advisory practice.

Building a Human Rights Program

Jonathan Drimmer

Abstract

In 2011, Jonathan Drimmer joined Barrick Gold Corporation, the world's largest gold mining company, with a remit to build a global human rights program—the first in the world under the UN Guiding Principles on Business and Human Rights, which were on the cusp of being adopted. Over the ensuing 8 years, Mr. Drimmer's colleagues at Barrick and he collaborated with a wide range of internal and external stakeholders, including Professor John Ruggie, and created a comprehensive program that relied on 4 core learnings, 5 key interpretive pillars, and 6 components. This "4 5 6" approach was honed over time. Numerous programs today emulate aspects of the program Mr. Drimmer and his colleagues created. Through this chapter, he shares key learnings gained through this work.

1 The Birth of a Human Rights Lawyer

I never actually had a first day at Barrick. It was March 2011. I had just left a secure and stable partnership at a major global law firm to join the world's largest gold mining company. I would be a senior lawyer there. I'd been at the firm for 7 years, following a stint focusing on war crimes-related issues as a Deputy Director of a unit in the criminal division of the U.S. Department of Justice.

Since arriving at the firm, I had been working at the intersection of anti-corruption and business and human rights. That meant providing advice to companies on how to develop processes to identify and prevent negative human rights impacts. The application of human rights principles to business still was an abstract concept. It was defined by disparate transnational tort cases—like the Alien Tort Statute—and

J. Drimmer (✉)
Washington, DC, USA
e-mail: jdrimmer@steptoe.com

A. McLaughlin, C. Chell (eds.), *Building the Case: How Lawyers are Leading on Sustainability*, https://doi.org/10.1007/978-3-031-88455-9_6

seemingly moving backwards with the failure of a UN Norms project that unsuccessfully tried to bring clarity to the space.

Back in 2004, businesses could sign on to the UN Global Compact's 10 principles. Those are a set of voluntary standards associated with responsible business conduct, and include human rights, as well as labor rights, environmental responsibilities and anti-corruption. They could also follow some sector-related principles, like the Voluntary Principles on Security and Human Rights. The Voluntary Principles focus on managing risks in the extractive sector. But human rights norms generally were considered applicable only to states. The notion that companies could and should take effective programmatic steps to mitigate negative impacts was embraced only by a handful of true believers.

The threat of high profile lawsuits and related campaigns began to seep into corporate psyches, however. There were sweatshop labor cases in the apparel sector in Asia. Ugly lawsuits detailed security abuses in oil, gas and mining. Headlines involving anti-union activities and child labor for agricultural companies splashed across newspapers. These caught the attention of corporate executives and boards. They saw peers named individually who did not want their reputations sullied. Companies began asking about measures to prevent human rights abuses. Taking a cue from anti-corruption compliance, I started giving them advice.

After about a year, I was still a fledgling business and human rights compliance lawyer, trying to figure out this emerging field. I'd been helping companies draft policies and trainings, provide oversight for programs, conducting investigations into incidents, and getting involved in the occasional transnational tort case. Through the grapevine, I heard about a professor from Harvard's Kennedy School, John Ruggie, who had been appointed as a new UN Special Representative. His goal: to develop principles regarding businesses and human rights.

I didn't think too much of it at the time. The Norms project had just imploded. I did not foresee an academic from a school of government to be the kind of visionary that would naturally lead global businesses through the desert to the promised land.

Over the next six years, after participating in consultations with John Ruggie and his team and providing information about human rights compliance approaches and transnational tort litigation, I became more hopeful.

2 The UNGPs

Unanimously endorsed by the UN Human Rights Council in 2011, the UN Guiding Principles on Business and Human Rights ("UNGPs") were transformative. They provided a coherent structure through three "pillars" applicable to states and business enterprises, reframing the social dimension of corporate responsibility in the language of duties and responsibilities.

The first pillar is the state duty to protect human rights. States have responsibilities over human rights-related issues within their jurisdiction by third parties, including business enterprises. As outlined by the UNGPs, it requires taking appropriate steps to prevent, investigate, punish and redress abuses through policies, legislation, regulations and adjudication. In essence, it restates international law.

The second pillar reflects the business responsibility to respect human rights. Pillar II represents Professor Ruggie's lasting contribution. It essentially translates international human rights law, previously conceived to apply to states, into a cogent framework for private actors. Business responsibility is defined by corporate connection to human rights impacts. Companies have responsibilities to respect human rights, avoid infringing on them, and address adverse impacts that occur according to a "cause, contribute and directly linked" formulation. The approach is defined by corporate governance, creating a practical compliance-oriented structure involving policies and procedures, training and education, assessments, internal and external reporting, and other key compliance concepts.

The third pillar is access to remedy. It includes multiple principles outlining how remediation should be considered for negative impacts that occur, introducing the concept of operational grievance mechanisms—conceptually taking the familiar concept of corporate hotlines and externalizing them to potentially impacted third parties and communities—and rights compatible remediation.

With the UNGPs, human rights programs were given a logical and practical outline—a skeleton on which a body could be built. It helpfully coincided with the basic practices of international regulatory compliance, providing support for the advice I had been providing for seven years.

3 The Adventure Begins

The year before the UNGPs were adopted, in Summer 2010, the international NGO Human Rights Watch approached Barrick and said there was a problem. Security guards at the company's mine in Papua New Guinea were gang raping illegal miners who came onto the property, Human Rights Watch reported. Still at the firm, I helped lead an extensive investigation into the allegations, and prepare the company's response—acknowledging the colorable evidence that emerged from the investigation, and pledging to take steps to make sure it never happened again. In early 2011, Barrick asked me to join them and create a program to do just that.

Which is how I found myself on a Qantas flight en route to Papua New Guinea on a Sunday night in March. It was just a few weeks after the final draft of the UNGPs was published. It also was a few hours from when I was to officially become a Barrick lawyer. Given flight times and international time changes as I traversed the globe, I landed on a Tuesday, never actually touching the ground on my Monday start date. Three months later, after my colleagues and I had outlined the contours of a human rights program and we were awaiting board approval of a new human rights policy, the UNGPs were adopted.

In hindsight, I was lucky in several respects. The adoption of the UNGPs, of course, were fundamental. They essentially endorsed the structured approach I'd considered vital to preventing negative human rights impacts. My colleagues at Barrick were outstanding—ambitious, innovative, and highly collaborative.

Though my real luck came a year later. In 2012, after Professor Ruggie's UN mandate ended, he joined Barrick's CSR Advisory Board. We could develop this comprehensive program, the first of its kind to follow the UNGPs, under the tutelage of the guy who wrote them.

4 Developing a Human Rights Program

The most important lesson I learned when I joined Barrick was the significance of systems and processes. They are the lifeblood of businesses. They create predictability, and common understandings. Systems and processes can be tested, measured and improved. They represent how work is done day-to-day in a way that is repeatable and sustainable. Human rights is no different. In the transnational regulatory and compliance space, these organized approaches to addressing corporate behaviors are called compliance programs. For other functions, they are called management systems.

The human rights program we created at Barrick was grounded in a basic management system philosophy. Originally, it was organized in 11 parts. Over time, we combined some of those components, calibrated it more closely to the structure of the UNGPs, and reorganized the program into 6 parts. The program matured and improved through trial and error, engagement with other companies and NGOs, and the advice and assistance of numerous internal and external experts, Professor Ruggie among them.

As is true of any program, the success of Barrick's was shaped by certain key principles that help guide thinking and behaviors. Most decent programs will have common practical features. The principles reflect the unique application of those features to the company and its particular culture and ethos. The principles are the difference between processes applied in a rote manner, and those embedded within the enterprise. The program also must be shaped by key learnings as it goes along. Improvements only come from humility and introspection, and integrating what you learn over time.

In rolling out the Barrick program, we came to settle on a "4 5 6" model. 4 key learnings stood out more than any others, shaping how we thought about the program in its application. They were integrated into 5 guiding pillars that animated the program's 6 structural elements.

4.1 The 4 Key Learnings

First, we learned that we could not take for granted that people will act ethically. The onus is on companies to hire people with integrity, to assess whether contractors, suppliers and customers accept and share their values, to develop systems and processes to prevent human rights violations, and to continually reinforce expectations and conduct training on how to implement those expectations. It takes persistent hard work and dedicated efforts to limit negative human rights impacts for any company, especially a multi-national enterprise like ours.

Second, we acknowledged that human rights dilemmas will persist. Certainly, having a comprehensive human rights program in place is an important start. But it is only a start and any human rights program will inevitably confront challenges that have no obvious and simple solutions, whether it is providing remedies for historical human rights violations, enabling access to potable water, or protecting workers from sexual harassment. We learned the importance of seeking creative answers, being willing to try new approaches that may not work, soliciting advice from those who have addressed similar situations, sharing our experiences with others, and engaging in collective action to pursue larger changes.

Third, we learned that we had to listen to a range of voices. This may have been the most difficult lesson. Our programs and activities generated a lot of comment, wide-ranging advice, and strongly expressed viewpoints from numerous actors—many of which were critical. We did not always agree with those comments and views, and fought the instinct to react defensively. But they compelled us to continue to closely examine our assumptions, behaviors and approaches, and determine whether they can be strengthened and improved. And just because someone is wrong 9 times in a row does not mean they are wrong the 10th. We learned to carefully consider the positions and assertions of all critical stakeholders, and not reject anything out of hand.

Finally, we learned that joint ownership is essential to operationalize a program. Key functional units associated with our salient risks, such as our security, environment, and safety and health groups, maintained detailed policies and procedures and management systems of their own. When implemented as designed, these protocols limited negative human rights impacts, though they did not expressly reference human rights. Embracing a global human rights approach meant reviewing how each functional unit around the globe may impact human rights, both positively and negatively. It meant adjusting procedures in light of human rights concerns, educating relevant personnel about how they may affect human rights, and vesting personnel with relevant accountabilities and responsibilities. We learned that respecting human rights is a shared global responsibility, belonging to no single person or group in the company.

4.2 The 5 Pillars of the Program

Informed by those learnings, along with the advice of Professor Ruggie and many others, our program included 5 key pillars that guided implementation.

First, it was grounded in global standards. The program was based on international human rights norms and guided by the UNGPs, the 10 principles of the UN Global Compact, the Principles on Security and Human Rights (VPs), and other similar standards. We treated human rights norms as obligatory and non-optional, like the laws that applied to the business. We came to believe that only through a globally integrated human rights compliance program could we meet our responsibility to respect the human rights of all our stakeholders.

Second, leadership from the top of the organization was critical. We believed that every employee in every functional unit played a role in respecting the human rights of our stakeholders. At the same time, leadership from the very top of the organization was essential to the program's success. That included vesting human rights oversight responsibilities with a committee of the board of directors and having executive management and local leaders reinforce the importance of respecting human rights.

Third, the program had to be embedded throughout the organization. We tried to integrate human rights considerations into the company's values, governance frameworks and the corporate management systems of different functional units. From supply chain and human resources to security and community relations, when human rights elements are entrenched within day-to-day job performance to a point that they are indistinguishable from other aspects of work activities, they are truly operationalized and ultimately sustainable.

Fourth, we believed in shared learnings. To improve our own practices and to advance business respect for human rights, we tried to share our successes, failures, and program features. We helped form, joined and led initiatives and working groups focused on dialogue and mutual learning, actively trying to advance human rights in the business community and offering examples of how it can be done and pitfalls to be avoided.

Finally, we actively sought out partnerships and collaborations. We realized that in facing human rights dilemmas that defy easy answers, successfully addressing them was only possible collectively with peers, business partners, and other stakeholders. At the same time, we tried to advance respect for human rights among the broader business community, helping others with the dilemmas they may have been facing.

4.3 The 6: The Elements of the Program

4.3.1 Policies and Procedures

Critical to an effective human rights program is a clear substantive policy and effective implementing documents. That is recognized in UNGP 16.

Among our first activities after my arrival was modifying our Code of Business Conduct and Ethics to include an explicit commitment to respecting human rights and creating a standalone Human Rights Policy. Both were adopted by the Board of Directors. While the Policy was periodically reviewed and adjusted, it contained the philosophical premise behind our human rights approach: that we would respect the human rights of all stakeholders impacted by our operations. It applied to all our employees, businesses and value chain. We specifically called out those human rights that created salient risks to the organization. We identified those international conventions and instruments, including the UNGPs, that helped define our practices. We also tried to make the policy simple and accessible, and created a series of

FAQs to help understand how it would be applied in practice. We had it translated into the local languages of the countries where we did business. We later refashioned it to a very simple statement of principles and commitments, a style that now is emulated throughout the industry.

Beyond creating a policy, we tried to embed human rights considerations into the company's values, governance frameworks and corporate management systems. The Policy was supported by and incorporated numerous function-specific policies, including those related to supply chain, human resources, EHS, security, community relations and others.

The tenets and precepts in the Human Rights Policy were buttressed by a number of dedicated policies and standards. These included a global internal procedure that required immediate reporting and escalation of information related to potential negative human rights impacts. Much as with anti-corruption risks, all information of potential human rights violations had to be reported, regardless of whether it seemed credible to the employee, or the amount of detail that the employee might know. Likewise, for external stakeholders, we believed that securing and maintaining a license to operate depended on our ability to listen actively and respond in a timely manner to issues of material importance. That included adopting grievance mechanisms at all operations, consistent with Pillar III of the UNGPs.

Another key procedure related to investigating human rights-related allegations. We required that all allegations of negative human rights impacts be investigated, though the nature and extent of the investigation would vary depending on the circumstances. Typically, for serious potential human rights breaches, we tried to create independence in our investigations.

As a further step, we created internal guidance documents to help guide employees in implementing our program. That included guidance on how to detect and respond to allegations of retaliation for whistleblowers, map employees for appropriate training, conduct due diligence on prospective employees and instituting relevant controls, and provide human rights remediation.

This comprehensive policy framework formed the foundational core on which the program was built.

4.3.2 Governance and Oversight

Governance reflects the structural management and oversight of a compliance program. It includes the role of the board, the responsibilities and accountabilities of management, and ways to help track a program's robustness and effectiveness. At Barrick, we vested day-to-day responsibility of the program in the Office of the General Counsel, where I sat. It was logical for several reasons. The General Counsel maintained a direct reporting line to relevant committees of the Board of Directors on compliance-related issues, including human rights and investigations. The General Counsel also oversaw our anti-corruption program and maintained authority over other international regulatory requirements at the company, helping create efficiencies with the human rights program.

Vesting the human rights program in Legal also was consistent with how we viewed human rights. We considered human rights norms to be legal obligations, and thus demanded adherence in the same way that we mandated compliance with other international and local regulatory requirements.

The Corporate Responsibility Committee of the Board of Directors, composed of four independent Board members, was charged with overseeing the human rights program. We provided the Committee quarterly updates on salient human rights risks, key issues, trends and projects. These included significant allegations of human rights violations, transparency and reporting efforts, investigations, remediation efforts, human rights assessment findings and outcomes, and collective action projects.

As part of the governance structure, the company's Corporate Social Responsibility Advisory Board—on which Professor Ruggie sat—also provided advice. Chaired by Barrick's President, the board convened twice per year, and as part of its mandate received regular updates on issues related to human rights matters at the company.

As the program was primarily implemented on a local basis, accountability for conduct rested primarily with our country leaders. Country legal heads were expected to provide local guidance, oversight and implementation of the program in their respective locales. At the same time, we undertook a risk-tiered approach, placing emphasis and resources on those locations and functions where human rights risks were most acute.

Critical to embedding human rights awareness across our operations was senior management visibly emphasizing the core principles covered by the program. This helped underscore the importance of the program, promoted acceptance, and helped create a culture where employees would strive to meet and not evade the requirements of the program. Indeed, while we had a robust training program, it was not possible to provide training on every potential situation that may arise. A culture of compliance helped employees abide by the values governing the program, and interpret events consistent with it.

We also took a range of other steps. We developed and tracked a range of operationally-relevant metrics. Once per month, at global Business Plan Reviews (BPRs) that included senior leaders of corporate functional units and sites, we reported on how the human rights program was progressing against its relevant goals and on issues that might prevent the program from achieving its objectives. Each site and advanced project also had to report on license to operate issues, which by definition included the human rights program. We also tied aspects of the program to executive compensation and our global bonus scorecard, which could impact employees across the company. For instance, one year, the bonus scorecard included key performance indicators related to messaging, employee onboarding, vendor due diligence and implementation of our security and human rights program.

4.3.3 Due Diligence

For any compliance program, it is important to conduct assurance to assess its effectiveness and strengthen processes found to have weaknesses. A human rights program is no different. UNGPs 17-21 recognize that. Given Barrick's size and diverse operational contexts, we realized that no single process could provide a full picture of the company's human rights impacts at any given location, much less around the world. We therefore took a holistic approach to understand our overall human rights footprint.

Human rights impact assessments: One of the cornerstones of our due-diligence efforts was a stand-alone, independent human rights assessment program for Barrick-operated properties. The assessments were conducted by a respected independent third-party consultancy, and focused on actual, potential and perceived impacts. A well-regarded NGO (Fund for Peace) served as an external independent advisor on the program, and Professor Ruggie also provided advice and guidance on discrete issues associated with the assessments. In-depth assessments of applicable country-level risks were developed by the Danish Institute for Human Rights.

The human rights assessment program began in 2011, and all high- and medium-risk sites were assessed repeatedly. The findings were incorporated into our overall assessment of enterprise risks under our enterprise risk management system, which were also discussed during the BPRs.

Notably, our approach—novel at the time—was distinguishable from lengthy one-time "stand-alone" assessments that companies occasionally pursued. It was also different from functional unit management system assessments, although we tried to create synergies with other assessments that relate to human rights to help to more effectively identify our human rights impacts and risks.

Internal and external audits: In addition to our human rights assessments, we conducted a variety of internal and external assurance activities related to the human rights program. The results were examined over a multi-year period to identify trends and changes. We also considered the findings in conjunction with information generated by other processes, such as through our enterprise risk management process, internal audits, grievances, hotline reports, our third-party annual social assurance process and community and stakeholder engagement programs, engagements with site and functional leads, and our investigations into incidents.

Employee screening: Important for any compliance program is screening potential employees whose background suggests they pose risks to the substantive principles at issue. At Barrick, all potential employees received pre-screening questions that could result in disqualification for employment based on past human rights violations, and background checks were conducted for employees in positions of trust. Employees in positions where human rights risks may be most prevalent, such as security, also received heightened vetting and due diligence. In addition, job applications and/or accepted employment letters and agreements contained statements expressing our human rights expectations.

Supplier screening and onboarding: We recognized, of course, that negative human rights impacts could occur as a result of behavior from our suppliers and their subcontractors. Therefore, as with potential employees, we educated potential suppliers on our human rights expectations before we entered relationships. We required that the supplier abide by Supplier Code of Ethics that incorporated the 10 Principles of the UN Global Compact and key concepts of our Human Rights Policy and Policy. We also included human rights considerations in our global Vendor Onboarding Standard. This includes basic due diligence on all direct suppliers related to human rights before contracting with them, and contemplated enhanced due diligence on suppliers with elevated risks of negative human rights impacts or who provided goods or services on-site.

Once Barrick agreed to do business with a supplier, the Code of Conduct and Human Rights Policy applied to them in their work for us, just as it did to our own employees. Our contracts with third parties also often had provisions requiring suppliers to abide by key human rights norms and Barrick policies, and we sometimes provided focused training to suppliers who have particular risks of negative human rights impacts. Under our Human Rights Policy, suppliers were expected to report human rights issues of which they become aware, and we would ask suppliers to periodically provide certifications that they were not aware of any unreported human rights allegations in their work.

4.3.4 Training and Advice

Educating relevant company personnel and third parties on company values, requirements, and expectations is a fundamental aspect of any compliance program or management system. The UNGPs specifically contemplate that, to embed a responsibility to respect human rights, business enterprises should create a human rights policy commitment that "is publicly available and communicated internally and externally to all personnel, business partners and other relevant parties." UNGP 16(d).

There were certain basic practices associated with training and education that we relied upon: (1) comprehensive training plans, beginning at induction and onboarding and with further trainings and reminders offered regularly thereafter; (2) training through a mix of formats, including live tailored sessions (particularly for individuals in positions of heightened risk or gatekeeping functions), online e-training, small group roundtables, "just-in-time" reminders associated with specific anticipated events and other behavioral prompts, as well as short "tool-box" updates or refreshers at the outset of meetings; (3) incorporating company values, using scenarios, hypotheticals, case studies, and lessons learned to help embed the company ethos, better enable application of key principles to situations that arise, and avoid policy violations; (4) providing targeted training for managers and supervisors in light of their positions in the company; (5) employing methods to evaluate the effectiveness of trainings; and (6) training in the local language of the audiences to facilitate comprehension. We tried to make training engaging and persuasive.

At Barrick, we began by conveying our human rights expectations before a new employee joined the company. The pre-screening for prospective employees helped us hire the best candidates, and made clear the primacy we placed on ethical colleagues. In addition, employment letters and agreements contained statements expressing our human rights expectations, again making clear that our human rights standards were unwavering. Upon joining the company, all of our employees, regardless of location or position, received basic human rights training during their induction. Certain relevant employees, including all security personnel, also received focused in-person human rights training pertaining to their areas.

On an annual basis, every employee with a computer received refresher training on human rights. Many other employees who did not have computers received training through other means, such as DVDs. We created a guidance to help sites identify employees who may be more exposed to certain human rights risks and who therefore should have received enhanced, in-person human rights training on an annual basis. We employed a "train the trainer" model to empower country-level employees to implement our training program and deliver relevant, accessible, and effective human rights training. For certain functional areas, such as security and community relations, we tracked the effectiveness of this training through internal assurance processes and other means.

To support training, we provided frequent internal messaging regarding the values associated with the program and compliance with the program through company newsletters, in dedicated human rights posters that were posted at all sites and projects, in flyers and handouts related to the program, via social media channels, and through other means. We also experimented with including theories of behavioral ethics in our trainings to try to drive employees to desired behaviors.

Finally, thousands of employees were required to sign annual certifications stating that they are not aware of any potential unreported violations of our Human Rights Policy, and agreeing to report any they may learn of. This also served as a reminder of the importance of the program.

4.3.5 Incentives, Disciplinary Action and Remedy

For compliance programs generally, we believed it was important to anticipate the corporate response when policies and procedures are breached, and incentivize good behavior. That included providing clear information to employees and third parties on the ramifications of breaching our Human Rights Policy, and other policies, as well as consideration of how remediation would be addressed. It also included identifying ways to reward positive behaviors.

Discipline. At Barrick, we tried to convey a clear message: violation of our Human Rights Policy and related procedures could (and did) lead to discipline, up to and including termination. That applied to avoiding human rights violations, and also to reporting information about known human rights violations and cooperating with investigators. Potential disciplinary measures ranged from individual counseling, to focused training, to oral or written warnings, to financial penalties, to termination in cases of serious breaches. In assessing disciplinary measures, a range of factors were considered, including: the significance of the procedural breach (and thus of the human rights violation), whether the individual received prior training in the area, the length of time the individual served in the position, whether the individual was on or off duty, the level of cooperation with investigators, showings of remorse and individual remediation efforts, whether the individual committed past breaches, and whether the breach was intentional.

We also applied our disciplinary processes to third party suppliers. And we were unequivocal in advising employees and third parties that, where we conclude that they have committed or contributed to serious human rights violations, we would cooperate with the police in prosecution efforts, and may assist victims in seeking redress directly against the perpetrators.

Incentives. At the same time, we tried to reward positive behaviors. We took steps to elevate ethical conduct alongside productivity in tangible ways. Consistently ethical behavior became part of performance evaluations and promotion decisions. Managers and supervisors were expected to lead by example and participate in trainings and roundtables. They were evaluated through anonymous "360 evaluations" that included questions related to ethics, compliance and integrity. We created ethics "scorecards" for managers in some locations, that included meeting compliance plans, tangibly demonstrating commitments to following and explaining relevant policies to other employees, visibly encouraging subordinates to speak up when they have questions or concerns, clearly communicating with subordinates about the importance of ethical decision making and the company's values, and any identified instances in which they did or did not allow business pressures to undermine ethics and compliance.

Grievance mechanisms & remedy. Barrick had internal global procedures that outlined the mechanisms that employees and third parties could use to notify local and corporate management of questions and concerns. These procedures also specified how we would investigate issues. Employees and third parties were routinely encouraged to use Barrick's Compliance Hotline to report any potential human rights violations they might see or hear about. We advertised broadly how to access the mechanisms, including on the website and intranet, on posters, on wallet cards, and through other means. We had a strong non-retaliation policy and developed an internal guidance document to assist supervisors in monitoring and responding to potential retaliation against those who made reports.

We also had policies and procedures so that every community where we operated had simple, accessible grievance mechanisms to provide feedback and request remediation for legitimate concerns. Grievance channels varied by site, and include message boxes, telephone hotlines, town hall meetings, supervisors (for employees), and direct contact with Barrick employees. Our annual human rights assessments included a review of how effectively the grievance mechanisms escalated potential human rights concerns, and the level of confidence in them. Our grievance mechanisms were also internally audited for implementation and effectiveness during regular assessments of our Community Relations Management System, and externally assessed against the UNGPs effectiveness criteria. We also reported publicly on our grievances and their resolution.

When we identified negative human rights impacts that we caused or contributed to, we tried to take culturally appropriate and thoughtful approaches to remediation and communication, and we tried to avoid obstructing access to other remedies in a manner consistent with the UNGPs. We also were sensitive to the importance of collaborative stakeholder and victim input in the development of appropriate remedy, and the potential need for independence in circumstances where remediation may have been appropriate. For serious negative human rights impacts, we developed corporate guidelines regarding the internal personnel involved in assessing how remediation of negative human rights impacts would be handled. We were particularly sensitive to the importance of victim participation, stakeholder input, as well as the potential need for independence from the operational unit that may have been involved in the negative impact. While remedies for negative human rights impacts naturally differed depending on the circumstances, in-kind remediation was often preferred to cash, and sites adopted guidelines that consider such factors, including: the degree and nature of the harm suffered, whether mine personnel were involved and on duty, whether third party perpetrators used mine resources or committed an act related to their contracted duties, the nature of the evidence in support of the claim, the individual's age and personal circumstances, and local laws. Where negative human rights impacts were caused or contributed to by entities in our value chain, we encouraged or otherwise tried to use leverage to facilitate their providing appropriate remedy and design processes to prevent recurrence.

4.3.6 Transparency

We tried to be transparent with external stakeholders, but like other companies, often had to balance competing risks associated with disclosures, including legal risks, potential harm to stakeholders, and confidentiality needs.

Nonetheless, consistent with the UNGPs, we publicized our human rights commitments to local communities and other stakeholders. This included through consultations with our Community Relations personnel, who we felt were best positioned to provide key information to potentially affected stakeholders in communities near our operations. As part of the VPs, we also engaged with and consulted local communities about security arrangements and our expectations around human rights. We tried to discuss openly our systems and processes for mitigating negative human rights impacts that we may have caused or contributed to with those potentially affected. We likewise tried to be open in our approach when negative impacts were caused by our suppliers, contractors and other third parties. And we sought to report formally when there were severe human rights impacts or risks of such impacts. We also directly contacted and engaged with key external stakeholders about human rights incidents, progress in remediation efforts, and the relative effectiveness of our systems and processes.

Much of our global external reporting occurred on our website, or through annual Responsibility Report or Human Rights Reports. Additional reporting was done at conferences, roundtables and workshops. Additional avenues of transparency also took place through partnerships with members of civil society. A report by UNICEF Canada on a pilot project regarding human rights due diligence at one of our mines was one example. Other examples included:

- Distributing booklets regarding human rights to local community members and government entities;
- Conducting community surveys on the prevalence of violence to gain feedback on how to improve security arrangements;
- Engaging public officials and community authorities on how to promote respect for law and order, reduce conflict, enhance and strengthen justice service delivery; and
- Providing education on the impacts of land purchase and sales.

5 Conclusion

I left Barrick in 2019. In some respects, I am back where I started—providing advice to companies on how to create and structure human rights programs. The advice I provide today is quite different from before going to Barrick. It has benefitted from the experience of actually developing and implementing a program in practice, seeing what works and what does not, and how to create sustainable systems and processes. It also benefitted from the wisdom of Professor Ruggie, and the collaboration with a range of other colleagues, experts and other stakeholders. I still talk about the theory behind a "4 5 6" approach, a structured program influenced by key learnings and principles that puts human rights aspirations into practice and confronts human rights dilemmas in the spirit of the UNGPs. I've been gratified to see many other companies take a similarly structured approach.

Looking back, I have my own key learnings from the experience. I learned to be bold; just because something in a program hasn't been done before does not mean it isn't a good idea or won't work. I learned it's ok to fail; when you try something new, it may not always work, at least the way you had hoped. Just don't let it happen too often. I learned to look for great people to collaborate with, and listen carefully to what they have to say. And as much as any thing, I learned that a bit of luck never hurts.

Jonathan Drimmer is a partner in Steptoe's Washington, DC office, where he resolves complex cross-border challenges with the benefit of having sat in every chair at the table: senior legal officer for a Global 500 company, federal prosecutor, and seasoned advocate. He is ranked Band 1 by *Chambers Global* for business and human rights law and is widely regarded as one of the world's leading authorities in the field. For more than two decades, he has advised and represented a wide range of companies and individuals on anti-corruption matters. His practice spans the full life cycle for both topics, including developing compliance programs, conducting investigations, and representing clients in enforcement proceedings and disputes.

Over the span of 30 years, Jonathan has directed hundreds of investigations worldwide, encompassing a vast span of geographies and substantive issues. He has represented companies and individuals in high-profile civil and criminal proceedings related to the Foreign Corrupt Practices Act (FCPA) and other anti-bribery laws, the Uyghur Forced Labor Prevention Act (UFPLA), the US Tariff Act and other laws banning forced labor imports, the Trafficking Victims Protections Reauthorization Act (TVPRA) and other anti-trafficking laws, the Alien Tort Statute, and similar matters. Jonathan has handled class actions in the US, Canada, and abroad, dealing with transnational torts, environmental cases, land controversies, and a variety of other similar areas. He has repeatedly represented companies and witnesses in congressional and parliamentary hearings related to human rights and anti-corruption issues.

Jonathan also has wide-ranging experience in helping companies manage multi-faceted dilemmas in overseas activities, developing and managing solutions that may encompass civil, criminal, regulatory, and political dimensions. He has designed a variety of human rights remedy programs and is particularly experienced in representing companies in non-judicial and quasi-judicial legal proceedings involving human rights and transnational torts. This includes disputes before the OECD National Contact Point disputes, claims to international human rights commissions, and inquiries by UN offices and agencies. He also has served as an expert witness in cases related to anti-corruption and business and human rights and has been appointed as an independent monitor or independent reviewer by companies and the UN on multiple occasions, particularly in relation to human rights and anti-corruption issues. Jonathan also has significant experience in assisting companies in developing best-in-class anti-corruption and human rights compliance programs, which include policies and procedures, training, due diligence, governance and all other program components.

He is a member of the US Federal Advisory Committee for Responsible Business Conduct, a senior advisor at BSR, the North American advisor to the Global Business Initiative on Business and Human Rights, and a strategic advisor to the secretariat for the Voluntary Principles on Security and Human Rights. Throughout his career, he has received numerous accolades and recognitions, including recognition as a Regulatory and Compliance Trailblazer and one of the 100 most influential in-house counsel in the US and Canada.

Previously, Jonathan served in the Justice Department as deputy director of the Criminal Division's Office of Special Investigations, where he led cross-border investigations, first-chaired prosecutions, and argued federal appeals. He was a partner at Am Law 100 law firms in Washington, DC, served as a Bristow Fellow in the Office of the US Solicitor General, and worked as a judicial clerk on the US Court of Appeals for the Ninth Circuit. His past board memberships include the Voluntary Principles on Security and Human Rights Initiative and TRACE International.

Jonathan was also the former deputy general counsel and chief compliance officer at Barrick Gold, one of the world's largest mining companies. He developed a compliance program at Barrick that has become an industry standard and has been replicated by numerous companies within and outside the extractive sector. Aspects of this program were the subject of a 2016 Harvard Business School case study.

Taming the Uncertainty: The Art & Science of Corporate Human Rights Counsel

Yousuf Aftab

Abstract

This chapter considers the proper role of counsel advising companies on how to navigate the suite of emerging corporate human rights laws across jurisdictions. The UN Guiding Principles on Business and Human Rights reshaped corporate responsibility from a public relations exercise to an objective, governance-focused discipline. That reconception has inspired a novel, dynamic, and interwoven web of regulation, litigation, and investor risk. Effective counsel in this arena depends on anticipating risk trends—both legal and non-legal; taming the uncertainty in risk substance and scope with principled definitions; and structured integration of priority risks in enterprise risk-management policies and protocols. The chapter explores the practical implications of legal trends and demonstrates in practice how effective counsel can derive resilient, precise, and structured governance from abstract human rights concepts.

1 Introduction

1.1 Corporate Responsibility Becomes Law

Business and human rights has evolved into a legal discipline at a breakneck pace. Not long ago, respected legal scholars could safely find that corporate responsibility "does not appear to fit comfortably within a traditional legal setting" because law endeavors to create "clarity and precision … [in] seeking a definition of key terms

Y. Aftab (✉)
Steptoe LLP, New York, NY, USA
e-mail: yaftab@steptoe.com

A. McLaughlin, C. Chell (eds.), *Building the Case: How Lawyers are Leading on Sustainability*, https://doi.org/10.1007/978-3-031-88455-9_7

and concepts or guidance on what constitutes acceptable forms of conduct".[1] Corporate responsibility as a concept intrinsically lacked such virtues: it reflected an *ideal* of ethical behavior embracing the interests of "stakeholders" beyond shareholders, but that ideal was subject to few shared metrics of right practice.[2] The ideal's breadth and vagueness made for a rhetorically powerful but substantively limited concept:

> [I]t means something, but not always the same thing, to everybody. To some it conveys the idea of legal responsibility or liability; to others, it means socially *responsible* behavior in an ethical sense; to still others, the meaning transmitted is that of "responsible for", in a causal mode; many simply equate it with a charitable contribution; some take it to mean socially conscious; many of those who embrace it most fervently see it as a mere synonym for "legitimacy", in the context of "belonging" or being proper or valid; a few see it as a sort of fiduciary duty imposing higher standards of behavior on businessmen than on citizens at large.[3]

From the mid-1980s, however, a series of "social" crises magnified attention on responsible corporate behavior.[4] The increasing recognition of corporate responsibility as a material business risk and opportunity inspired attempts to create more structured standards for business in the form of voluntary codes and law.[5] These standards culminated in the UN Guiding Principles on Business and Human Rights (Guiding Principles),[6] which have become the cornerstone of corporate responsibility as a legal discipline.

Unanimously endorsed by the UN Human Rights Council in 2011, the Guiding Principles have since been widely embraced by governments, industry associations, businesses, international organizations, and bar associations.[7] In recent years, they have been integrated into regulatory frameworks—notably the European Union Corporate Sustainability Due Diligence Directive (CS3D)[8]—and major judicial decisions across jurisdictions.[9] The Guiding Principles reframe the social dimension of corporate responsibility in the language of rights. They are built on three "pillars," conceived for the Guiding Principles to apply comprehensively to all states and business enterprises.

The Guiding Principles' novel contribution was a practical framework to understand the scope of business responsibility for human rights *as distinct from* the state responsibility for rights. While states are expected to "protect" human rights,

---

[1] Kerr et al. (2009), p. 5.

[2] Kerr et al. (2009), pp. 6–7.

[3] *Id.* (alteration in original) (quoting Votaw 1973).

[4] Albareda (2008), p. 432; Spence (2011), pp. 72–73.

[5] Zalik (2015).

[6] Ruggie (2011), adopted by the UN Human Rights Council on June 16, 2011.

[7] *See* Aftab (2014), § 19.02; *see also* Sherman (2012), p. 6.

[8] European Parliament and Council Directive on Corporate Sustainability Due Diligence 2024/1760, 2024 O.J. (L) [hereafter "CS3D"].

[9] *See* Milieudefensie et al. v. Royal Dutch Shell plc, No. C/09/571932 (Hague Dist. Ct. May 26, 2021).

businesses are responsible for "respecting" them. Respect is defined by corporate governance, broadly comprised of: (1) a policy commitment; (2) a due diligence process; and (3) a remediation process (including grievance mechanisms).[10] The core of this system is due diligence, which has a legally idiosyncratic meaning, to capture steps from assessment to disclosure. "The process should include assessing actual and potential human rights impacts, integrating and acting upon the findings, tracking responses, and communicating how impacts are addressed."[11]

The governance-based understanding of responsible corporate behavior marked a fundamental transition of corporate responsibility from elusive, reputation-focused pursuit to a precise, harder-edged legal and investor discipline. In recent years, governments have attempted different routes to regulate business-related human rights impacts across borders. The most ambitious laws—such as Germany's Supply Chain Due Diligence Act (LkSG),[12] Norway's Transparency Act,[13] and the French Loi de Vigilance[14]—mandate due diligence covering a broad array of human rights and reaching several tiers deep into corporate supply chains. Even more expansive legislation will flow from the CS3D,[15] enacted earlier this year. Other jurisdictions have adopted narrower laws, focusing on modern slavery or child labor, or more limited governance expectations, largely centered on disclosure.[16] Human rights-related trade sanctions have also gained ground in the United States and Europe, representing a blunter form of regulation.[17] Each of these regulations

[10] Guiding Principles, GP 15.

[11] *Id.*, GP 17.

[12] Leiferkettensorgfaltspflichtengesetz [LkSG] [Act on Corporate Due Diligence Obligations in Supply Chains], July 16, 2021, Bundesgesetzblatt, Teil I [BGBL. I] at 2959 (Ger.) (requiring German companies of more than 3000 employees to prevent or minimize human rights and environmental risks in their supply chain) [hereafter "LkSG"].

[13] Lov om virksomheters åpenhet og arbeid med grunnleggende menneskerettigheter og anstendige arbeidsforhold (åpenhetsloven) [Act relating to enterprises' transparency and work on fundamental human rights and decent working conditions (Transparency Act)], July 1, 2022, Lovvedtak 176 (2020–2021) (Nor.) (requiring companies to carry out due diligence in accordance with the OECD guidelines).

[14] Loi 2017-399 du 27 mars 2017 relative au devoir de vigilance des sociétés mères et des entreprises donneuses d'ordre [Law 2017-399 of Mar. 27, 2017 on the Duty of Vigilance of Parent Companies and Ordering Companies], Journal Officiel de la République Française [J.O.] [Official Gazette of France] Mar. 28, 2017 (requiring French companies to publish public vigilance plans annually that outline steps taken to identify and prevent adverse human rights and environmental impacts resulting from their own activities, the activities of companies they control, and the activities of their subcontractors and suppliers with whom they have an "established commercial relationship).

[15] CS3D supra note 8.

[16] *See, e.g.*, Modern Slavery Act 2015, c. 30 (UK); Modern Slavery Act 2018 (Cth) (Austl.); An Act to enact the Fighting Against Forced Labour and Child Labour in Supply Chains Act and to amend the Customs Tariff, Bill S-211, 1st Sess., 44th Parl., 2022 (Can.); California Transparency in Supply Chains Act, Cal. Civ. Code § 1714.43 (West 2019).

[17] *See* Uyghur Forced Labor Prevention Act, Pub. L. No. 117-78, 135 Stat. 1525 (2021); *Commission Proposal for a Regulation of the European Parliament and of the Council on Prohibiting Products Made with Forced Labour on the Union Market*, COM (2022) 453 final (Sept. 14, 2022); Regulation 2017/821, of the European Parliament and of the Council of 17 May 2017 Laying Down Supply

advance—expressly or implicitly—the Guiding Principles' governance-based model of corporate respect for human rights.

1.2 My Sustainability Career

The emergence of the Guiding Principles marked the beginning of my career in corporate sustainability. Until that point, I had been a corporate-responsibility skeptic. It appeared to me to be a marketing exercise—one that was possibly misguided from a business perspective. One that was not substantive, principled, or certain enough to ground a legal analysis or discipline. Indeed, I was even skeptical of the power of international human rights law to drive substantive progress; its aims and principles were clear and good, but respect for them seemed be beholden to government will and power, both too often inconstant when needed most.

That perspective started to change in my second year of practice at Latham & Watkins LLP in New York. I was a litigation associate focusing on international arbitration when a pro bono client, an international institution, asked us to explore the implications of an emerging corporate responsibility framework proposed by Professor John Ruggie, the UN Secretary-General's Special Representative on human rights and transnational corporations. The "Protect, Respect, and Remedy Framework," which later became the Guiding Principles, was both intellectually and practically fascinating to me. Intellectually, I was drawn by the framework's central challenge of coherently and practically translating public international law concepts—human rights—into private, extra-legal responsibilities of multinational companies. Practically, I was drawn to the promise of diversifying the responsibility to achieve the ends of international human rights beyond states to corporations.

Despite this fascination, however, I remained apprehensive regarding the likelihood of voluntary standards—as the Ruggie framework was conceived to be—to drive meaningful, systemic change in the conduct of business or in lives of vulnerable individuals and groups the world over. That apprehension would lessen and then turn to true optimism in the years to come. The lessening happened as I observed global businesses from across sectors, many of them our clients, start to publicly support the framework, which suggested a greater appetite for normative corporate responsibility guidance from the private sector than I had anticipated. The complete turn happened at the instigation of an investor-state arbitral tribunal that asked about the meaning of a term of art in international law, "in accordance with law," which conditions investor access to the protections of bilateral investment treaties.

In preparing the submission, I researched the legal evolution of international corruption standards and came to see a possible analogue to the coming legal evolution of corporate human rights standards. My hypothesis was that, even if they remained voluntary, the Guiding Principles could ground increasing legal liability for global

Chain Due Diligence Obligations for Union Importers of Tin, Tantalum and Tungsten, Their ores, and Gold Originating from Conflict-Affected and High-Risk Areas, 2017 O.J. (L 130) 1.

companies either as representation or in reshaping the standards of the reasonable business. The more I spoke to our clients about what they were doing to implement the Guiding Principles, however, the more I realized that the fundamental shift those standards represented had not been fully grasped by the companies likely to face the greatest risk. Most of those clients did not appreciate that the emergence of the Guiding Principles had reshaped corporate responsibility from a communications discipline to an objective, substantive, and standard-based one.

The dissonance between what I believed were the risk implications of the Guiding Principles and the extant method for managing voluntary corporate commitments to them created a gap in the market. Most sustainability practitioners, in-house and external, were treating the Guiding Principles as if their words had no objectively ascertainable meaning or non-reputational consequence. Corporate sustainability programs thus remained largely as they were, oblivious to the increasingly legal content and consequence of the soon-dominant corporate responsibility standards. I launched my first corporate human rights strategy firm, Enodo Rights, in 2012 to address this gap by leveraging legal expertise to offer structured advice tailored to emerging legal risk.

1.3 Key Legal Risk Trends

Just a decade after the launch of Enodo, the global legal landscape governing corporate human rights responsibility is complex, multidimensional, and evolving. Existing and emerging laws take materially different paths to regulating corporate management of extraterritorial human rights risks. Those divergent approaches have distinct implications for the structure of corporate compliance programs and the corresponding legal risks companies face. Nonetheless, they reflect six key trends for legally resilient corporate human rights programs.

1.3.1 Human Rights Risks Cannot Be Siloed from Other Sustainability Risks

Relevant risks for companies to manage are increasingly broad, spanning environmental and social issues. Germany's LkSG, for instance, weaves together environmental and human rights risk management in general. One of the specific "human rights risks" is expressly an environmental one, focused on the health and livelihood impacts of pollution and resource use.[18] So too the CS3D, which integrates under the concept of relevant "adverse impact" the full array of human rights and environmental issues.[19] It is thus increasingly difficult to separate "human rights risks" from broader non-financial risks to the company and to stakeholders.

[18] LkSG, *supra* note 12, §2(2)(9).

[19] CS3D *supra* note 8, annex I.

1.3.2 Relevant Risks Span the Entire Value Chain, Vertically and Horizontally

Relevant human rights risks are not confined to upstream issues, let alone to first-tier suppliers. While there are nuances in the scope of different laws, the trend is for due diligence and remediation responsibilities to extend through sub-tier suppliers all the way to the extraction of minerals. The LkSG also takes pains to emphasize corporate human rights responsibility in its "own business area" and "across all relevant business processes."[20] So too the CS3D, which extends these responsibilities downstream as well.[21] To ensure regulatory compliance and address disclosure risks, companies subject to such legislation will need to apply a human rights lens broadly to an array of functions from human resources to research and development, privacy, government relations, and marketing.

1.3.3 Knowledge May Create Responsibility and Liability

The Guiding Principles were structured to encourage due diligence, in part by defining corporate responsibility for human rights independently of knowledge.[22] Under emerging legislation, however, knowledge can itself increase liability. Thus, for instance, under the LkSG, a company's responsibility for impacts of indirect suppliers increases based on knowledge. While companies do not need to conduct risk assessments of indirect suppliers, if they obtain "factual indications" of risks at such suppliers, companies must develop a concrete plan to address the risks.[23] Similarly, under the Norwegian Transparency Act, a company that is aware of human rights impacts about which information is sought may have fewer grounds—if any—to resist disclosure at the request of interested parties.[24] Companies may thus need to calibrate diligence to the implications of additional knowledge, particularly in their sub-tier supply chains.

1.3.4 Risks Should Be Prioritized Based on Salience

One of the key implications of these new laws is to shift the perspective for prioritization. Rather than materiality to the business alone, companies are increasingly expected to prioritize risks based on severity of risk to stakeholders, i.e., "salience." Salient human rights issues reflect the most substantial potential negative impacts of the company's activities or business relationships, regardless of the impact on the business. Salience is premised on two broad elements: (1) severity, in terms of how

[20] LkSG, *supra n*ote 12, §§4-7.

[21] See, e.g., European Parliament Resolution on corporate due diligence, Annex, §3(5) (defining a company's "value chain" across which due diligence obligations apply as "all activities, operations, business relationships and investment chains of an undertaking and includes entities with which the undertaking has a direct or indirect business relationship, upstream and downstream.").

[22] Guiding Principles, GP 22 Commentary (emphasizing that "cause" and "contribute" do not depend on foreseeability).

[23] LkSG, s*upra* note 12, §9(3).

[24] Transparency Act, *supra* note 13, §6 ("The right to information regarding actual adverse impacts on fundamental human rights with which the enterprise is familiar, applies irrespective of the limitations in the second paragraph."); see also Commentary, 8.3.3.4.

grave and widespread the impact might be on those affected; and (2) likelihood of harm.[25] The privileging of salience in law aligns with the concepts of "dynamic materiality" and "double materiality," which are gaining increasing traction in investor literature and in disclosure legislation.[26] This requirement could have radical implications for corporate governance of human rights risks.

1.3.5 Material Uncertainty Will Endure in Governance Scope and Expectations

The Guiding Principles are built on novel and uncertain legal concepts, notably the concept of 'adverse impact', discussed below. This uncertainty is manageable with voluntary expectations. The situation changes markedly when legal sanctions attach to purported non-compliance with the Guiding Principles' duty of care. It will be particularly challenging to anticipate the precise contours of laws like Norway's Transparency Act, which incorporate voluntary standards by reference, both in substantive scope and specific governance responsibilities. But lacunae will likely exist even with more precise laws like the LkSG because of their breadth. This challenge is all the greater with the CS3D, which incorporates a suite of international legal instruments tailored to governments, not businesses. The uncertainty is likely to create opportunities for stakeholder criticism and regulatory oversight that will require particular attention by companies to tame preemptively.

1.3.6 Business Risks Are Interdependent and May Be at Odds Across Jurisdictions

A uniting thread in each of these emerging pieces of legislation is the aim of leveraging disclosure to drive better corporate behavior. Norway's Transparency Act is expressly designed to leverage investor, civil society, and consumer pressure.[27] So too the EU Corporate Sustainability Reporting Directive, which, when combined with the CS3D, creates detailed and comprehensive disclosure and due diligence obligations for subject companies. While the German LkSG's oversight approach is more prescriptive and administrative, it similarly makes ongoing disclosure a core requirement. These expectations are consistent with the California Transparency in Supply Chains Act and Modern Slavery Acts from the UK and Australia. Beyond the reputational and investor pressures the disclosure requirements create, disclosure can also create risks under various legal regimes, from voluntary assumption of parent responsibility in transnational tort cases,[28] to material misrepresentation

[25] *See* UN Guiding Principles Reporting Framework, "Salient Human Rights Issues," online: https://www.ungpreporting.org/resources/salient-human-rights-issues/.

[26] See, e.g., White Paper, World Econ. Forum & Boston Consulting Grp., "Embracing the New Age of Materiality: Harnessing the Pace of Change in ESG" (Mar. 2020), online: https://www.weforum.org/whitepapers/embracing-the-new-age-of-materiality-harnessing-the-pace-of-change-in-esg.

[27] Transparency Act Proposition at 8.3.3.

[28] Such cases are particularly notable in Canada and the UK: *Choc v. Hudbay Minerals Inc.*, [2013] O.J. №3375, 2013 ONSC 1414, 116 O.R. (3d) 674 (Ont. S.C.J.); *Araya v. Nevsun Resources Ltd.*, [2017] B.C.J. №2318, 419 D.L.R. (4th) 631 (B.C.C.A.); *Garcia v. Tahoe Resources Inc.*, [2017]

claims by investors or consumers,[29] to trade sanctions.[30] It will therefore be critical for companies to ensure that compliance in one jurisdiction, notably with disclosure requirements, does not create material risk in another.

1.4 Practical Governance Implications

Against the backdrop of proliferating and materially distinct legislation across jurisdictions, global companies risk wasting time and resources by focusing only on achieving perfect compliance with existing legislation. Rather, the pursuit should be a risk-management program that is nimble and resilient with a view to emerging trends, to enable rapid adaptation to shifting requirements. That is, companies should seek to establish parameters for human rights governance designed to navigate the six trends above while efficiently leveraging the company's existing compliance infrastructure. This end can be reasonably pursued in many ways to align with corporate organization and culture. Three key considerations should animate the structure of efficient, forward-looking human rights governance against a fluid and evolving legal landscape.

1.4.1 Human Rights Risk Management Is Too Big for One Team

Meeting the compliance requirements of emerging regulations will require risk-management measures to be embedded across all functions and business divisions, from procurement and partnerships to human resources, marketing & sales, and government relations. No one team, irrespective of expertise or resources, can be tasked with designing and implementing the company's human rights strategy.

Rather, corporate human rights responsibilities are better divided into four broad categories: (i) *strategy* to navigate an array of human rights-related risks and opportunities; (ii) *implementation* of due diligence and remediation protocols across relevant functions, business lines, and value chain; (iii) *engagement* with key stakeholders, including worker associations, civil society groups, investors, and regulators; and (iv) *disclosure* to meet regulatory, investor, and key stakeholder expectations.

B.C.J. №117, 2017 BCCA 39 (B.C.C.A.); *Das v. George Weston Lt.*, [2017] O.J. №3542 (Ont. S.C.); *Chandler v. Cape plc.*, [2012] EWCA Civ 525, [2012] 3 All E.R. 640; *AAA and others v. Unilever plc and another company* [2017] EWHC 371, [2017] All E.R. (D) 07 (Q.B.); *Lungowe and others v. Vedanta Resources plc and another*, [2017] EWCA Civ 1528, [2017] All E.R. (D) 102; *Okpabi and others v. Royal Dutch Shell plc and another*, [2018] EWCA Civ 191, [2018] Bus. L.R. 1022.

[29] Misrepresentation cases have been most frequently brought in California, with limited success to date: *Kasky v. Nike, Inc.* 45 P.3d 243 (Cal. 2002), cert. denied 539 U.S. 654 (2003); *Dana v. Hershey Co.*, 180 F. Supp. 3d 652, 2016 U.S. Dist. LEXIS 41594 (N.D. Cal. March 29, 2016); *Sud v. Costco Wholesale Corp.*, 2016 U.S. Dist. LEXIS 5524 (N.D. Cal. January 15, 2016); *Wirth v. Mars Inc.*, 2016 U.S. Dist. LEXIS 14552 (C.D. Cal. February 5, 2016); *Barber v. Nestlé USA, Inc.*, 154 F. Supp. 3d 954, 2015 U.S. Dist. LEXIS 170608 (C.D. Cal. December 9, 2015).

[30] US Tariff Act 1930; Uyghur Forced Labor Prevention Act, 135 STAT. 1505 (2021).

Human rights strategy, engagement, and disclosure likely can and should be managed by a dedicated, expert team to ensure coherence. But effective and efficient implementation will require a cascading of responsibility throughout the organization, built to track existing reporting lines.

1.4.2 Counsel's Oversight Role Is Critical

Legal counsel is essential to effective human rights risk management to tame material uncertainty, ensure compliance, and manage information flow. *First*, the inherent uncertainty in emerging legislation will require proactive management through coherent definitions of core terms, as illustrated in the next section. *Second*, the risk of regulatory sanction will mean that decisions on human rights risk identification and response will need to be increasingly precise—particularly where regulators do not afford deference to business judgment. Counsel will be essential in structuring the human rights process flow and decision structure to ensure internal consistency and external coherence with regulatory requirements. *Third*, counsel will be essential to manage information flow securely to navigate the potentially conflicting incentives of distinct regimes, e.g., when disclosure under the Norwegian Transparency Act may create the risk of trade-related sanctions in the United States. That role includes preserving privilege over sensitive due diligence and balancing the benefits of rigorous assessments against the additional responsibility for knowledge.

1.4.3 Human Rights Governance Cannot Be Exclusively Focused on Legal Compliance

Despite the proliferation of legislation, corporate human rights strategy cannot be tailored just to navigate legal risks. Human rights risks are particularly challenging for companies precisely because material issues span brand, investor, legal, political, and operational risks. All these risk dimensions are interdependent, and two or more will frequently be material at the overarching governance level and with respect to any specific issue. Moreover, the counterpoint of many human rights risks is opportunity to strengthen the brand or relations with investors and clients. An effective human rights strategy and program will constantly pay heed to and weigh each of these elements of business risk and opportunity. To that end, cross-functional expertise including legal, government relations, communications, procurement, and sales is essential to avoid a technocratic and formalistic approach to human rights risk management.

2 Taming the Uncertainty: The Role of Counsel

The fascinating element of this area of practice is that the role of counsel is evolving—and must keep evolving—in tandem with the substantive evolution of the law itself. The emergence of business and human rights as a legal discipline is, at a high level, the chiseling of legal certainty into an amorphous moral inclination. That pursuit of certainty has two related dimensions: identifying relevant risk and building effective risk management. Underlying the whole is a question about the nature of business itself, and the proper place of conventional and well-established corporate pursuits against the backdrop of increasingly stringent expectations of companies to guard, if not advance, the public interest. That is: how can leaders of a company respect human rights while acting in accordance with their fiduciary duty?

The answer to that question shapes and will be shaped by how this discipline evolves. For now, however, corporate counsel's role is increasingly to help tame the uncertainty in substance and strategy, against a dynamic risk backdrop. Given these pressures and uncertainties, it is increasingly critical for corporate counsel to wrestle with the proper balance of business risk and opportunity and stakeholder risk and opportunity. Two elements of this ongoing pursuit bring to the fore particularly legal expertise. The first concerns the scope of business responsibility. The second concerns the manner of integrating human rights in broader enterprise risk management (ERM). I summarize the challenges below before drawing on an engagement to illustrate how practically to navigate them.

2.1 Challenge 1: Defining Risk

The Guiding Principles expect businesses to "respect human rights," i.e., to "avoid infringing on the human rights of others" and "address *adverse human rights impacts* with which they are *involved*."[31] An "adverse impact" is one that "removes or reduces the ability of an individual or group" to enjoy an internationally recognized human right.[32] This concept is the Rosetta Stone of the Guiding Principles and emerging law: it differentiates corporate from state human rights responsibility (adverse impact v. abuse, violation, or infringement); sets the aim of due diligence; and establishes the parameters of expected remedy.

The key certainty challenge is that human rights under international law have traditionally been *state-centric* concepts. The practical implication of this challenge is that international human rights, as legal concepts, are defined through prohibitions and requirements of states.[33] The certainty challenge is threefold. *First*, for most international human rights, there is no law establishing the legal responsibilities of business as distinct from states. *Second*, and more problematically, there is no authoritative *method* to translate human rights instruments into the private context. *Third*, when is an impact on a right "adverse"? International law distinguishes between *legitimate* and *illegitimate* interference with human rights.[34] This

[31] Ruggie (2011), emphasis added. Notably, this differs from a State's obligation under international human rights law, which is to "respect, protect and fulfill" human rights.

[32] U.N. Off. High Comm'r for Hum. Rts., *The Corporate Responsibility to Respect Human Rights: An Interpretive Guide*, at 5, U.N. Doc. HR/PUB/12/02 (2012) ("an action [or omission] removes or reduces the ability of an individual to enjoy his or her human rights").

[33] Pati (2005), pp. 242–243.

[34] Pati (2005), p. 225; Aftab and Mocle (2019).

distinction is critical to understanding what a right is and when it has been violated. The result is that, in many cases, it is very difficult for a company to reasonably foresee when it might "adversely impact" a human right.

2.2 Challenge 2: Integrating Risk Management

A second critical challenge for counsel wrestling with managing increased corporate human rights expectations is how to operationalize these definitions in practical, effective, and efficient risk-management programs. The challenge is that, unlike other risk areas like corruption or health and safety, the scope of relevant human rights risks can appear overwhelmingly vast. Under the Guiding Principles, these risks span the entire value chain and all corporate functions. It would be impossible for any one group or function to manage this arena alone. Indeed, the express expectation in the CS3D and the LkSG is that companies integrate human rights expectations across functional policies and enterprise-wide risk management.

Integration can be achieved in different ways. As with much of this discipline, it is as much art as science. No matter the approach, the key elements are precision and structure. Precision relates to (i) the definition of the relevant risk; (ii) the differentiated levels of possible company involvement with the defined risk; and (iii) the level of the risk's salience and materiality, which shape its priority under relevant law. Structure relates to proper categorization of the risk type and effective distribution of risk-management responsibility. That is, to be meaningfully embedded at the enterprise level, human rights risks need to be classified in a way that enables tailored and practical management by those with the right level of authority and functional expertise.

To illustrate what effective advice might look like in practice, I draw on an amalgam of engagements we have had with resource and energy (R&E) companies regarding the intersection of environmental and human rights risks. Resolving the first challenge—defining the risk in a way that is legally defensible—requires principled creativity, deriving risk parameters from international law in a way that reasonably apply to private actors. The second challenge involves leveraging and refining existing risk-management protocols to capture the scope and dimensions of the defined risk. The combination creates the structure necessary for precise and practical integration that will be resilient to future scrutiny by regulators, courts, and even reasonable civil society actors.

3 Lessons from Practice

The most relevant right to consider when framing an R&E company's environmental responsibility through a human rights lens is the right to health, enunciated in several instruments of international law. The cornerstone of the right to health under international law is Article 12 of the International Covenant on Economic, Social and Cultural Rights (ICESCR), which affirms "the right of everyone to the enjoyment of the highest attainable standard of physical and mental health."[35] Article 12 enumerates four types of measures expected of states to realize the right:

1. Measures to ensure healthy birth and child development.
2. Measures to improve environmental and industrial hygiene.
3. Measures to prevent, treat and control disease.
4. Measures to "assure to all medical service and medical attention in the event of sickness."[36]

The element most relevant to the R&E sector concerns measures to improve environmental and industrial hygiene. State responsibilities under this dimension include measures to prevent occupational accidents and diseases; ensuring a sufficient source of "safe and potable water and basic sanitation"; prevention and reduction of exposure to harmful substances that impact health; minimizing causes of occupational health hazards; adequate housing and safe and clean working environments; sufficient supply of food and adequate nutrition; and discouraging the abuse of alcohol, drugs, and other harmful substances.[37]

For this right to be meaningful in determining business-specific responsibility, however, it would need to be translated from a public to private context—recognizing the distinct institutional roles and powers of business versus states. That transposition is the cornerstone of a resilient risk-management system because it underpins consistent risk triggers and responses across the company (in the same way that clear definitions of corruption are critical to a coherent compliance system). As human rights are inherently legal concepts—defined in legal instruments,

[35] *International Covenant on Economic, Social and Cultural Rights*, adopted 16 December 1966, entered into force, 3 January 1976, UNTS 993, art. 12. Numerous international human rights treaties have codified the right to health. Article 5(e)(iv) of the International Convention on the Elimination of All Forms of Racial Discrimination requires States to prohibit discrimination in "public health, medical care, social security and social services." *International Convention on the Elimination of All Forms of Racial Discrimination*, adopted 21 December 1965, entered into force 4 January 1969, 660 UNTS 195, art. 5(e)(iv). Article 12 of the Convention on the Elimination of All Forms of Discrimination against Women requires that women receive equal access to health care and that they also receive pregnancy-related care. *Convention on the Elimination of All Forms of Discrimination against Women*, adopted 18 December 1979, entered into force 3 September 1981, 1249 UNTS 13, art. 12. The Convention on the Rights of the Child contains many protections including, for example, guarantees for necessary pediatric care and access to information on pediatric health and nutrition. *Convention on the Rights of the Child*, adopted 20 November 1989, entered into force 2 September 1990, 1577 UNTS 3, art. 24. Other human rights instruments guarantee the right to health in specific situations such as armed conflict, development, and the workplace. Paul Hunt, Report of the Special Rapporteur on the right of everyone to the enjoyment of the highest attainable standard of physical and mental health, 13 February 2003, E/CN.4/2003/58, ¶ 12 (citations omitted).

[36] Id.

[37] U.N. Comm. on Econ. Soc. and Cultural Rts., *General Comment No. 14 (2000): The right to the highest attainable standard of health (article 12 of the International Covenant on Economic, Social and Cultural Rights)*, ¶ 15, U.N. Doc. E/C.12/2000/4 (Aug. 11, 2000) [hereinafter General Comment 14].

jurisprudence, and commentary—this transposition exercise is necessarily the preserve of counsel.

I provide a sample transposition below. While the full reasoning is beyond the scope of this chapter to illuminate, it is based on authoritative international jurisprudence and commentary, filtered through the prism of business's legal responsibilities, and adapted to the innate institutional differences between states and businesses. *NB*: This is a justifiable definition, but not the exclusive one, as there is no authoritative jurisprudence canvassing the meaning of the right in this context.

Right to Health: Environment

(1) An R&E company adversely impacts the right to health: environment if it materially limits or impedes government efforts, through progressive realization to the maximum of available resources, to fulfill the right of everyone to the enjoyment of the highest attainable standard of physical and mental health through measures to improve environmental and industrial hygiene.
(2) An R&E company adversely impacts the right to health: environment if it materially harms individual, group, or community mental or physical health through acts or omissions that substantially diminish the quality of environmental and industrial hygiene.
(3) An R&E company adversely impacts the right to health: environment if it fails to meet a legal, contractual, or other legitimate expectation[38] to implement measures that would materially improve the quality of environmental and industrial hygiene.
(4) Nothing in this right obliges R&E companies to breach local law or to engage in governmental functions, unless mandated by law, contract, or other legitimate expectations. R&E companies may legitimately restrict this right to the extent lawful and reasonably justifiable to meet their fiduciary duties to shareholders and other stakeholders, including the reasonable exercise of legal rights to secure, protect, and profit from natural resources.

This is a particularly complex issue area that we have used for illustrative purposes. But it is powerful to demonstrate the kind of reasoning that will be necessary for a resilient program that anticipates future scrutiny by regulators, courts, investors, and stakeholders alike. While we await the essential certainty that would enable more predictable compliance, it will be incumbent on companies and industry associations to develop coherent, practical, and justifiable definitions of these core terms reasoning from first principles and adapted to the context.

Enterprise Risk Management Integration

The definition above provides only the first step in building a coherent and resilient risk-management system. The next step is to integrate the risk parameters in

[38] "Legitimate expectation" comes from the jurisprudence regarding potential omission liability. It would be a context specific analysis to cover an array of factors such as prior representations or special relationships that might ground legitimate government, individual, and group expectations of a special duty.

appropriate governance for ongoing assessment and response. I share one approach our team implemented when refining an R&E company's ERM program to better integrate human rights on an ongoing basis.

We were impressed with the breadth and rigor of the company's ERM program, which comprehensively and precisely tracked a range of priority financial and non-financial risks. When we began our engagement, the company was ahead of many peers in that human rights were already part of its ERM program. But they were categorized as a reputational risk under one overarching heading. This classification was both misleading and difficult to operationalize. It was misleading to the extent human rights were treated as largely a brand issue, rather than a legal or operational risk. Moreover, we found that a single category may not allow for segregation and precise human rights analysis for the purposes of an effective risk assessment, as human rights risks run across the company's value chain, with different practical implications based on the company's potential involvement.

We therefore drew on the ERM's operational risk categories to ground a functional-operational map of the distinct areas of the company's business in each division that might impact human rights in different ways. The final map in Excel was divided into five columns (each with sub-columns):

i. **ERM Value Chain Category**: We used this referent to ensure that human rights risks are not siloed from the company's broader operational risks.
   a. Exploration
   b. Development
   c. Procurement
   d. Production
   e. Marketing & Distribution
   f. Human Resources
   g. Information Technology
   h. Finance & Tax
   i. Government Relations
ii. **Stakeholder Risk**: We disaggregated potential risks from the perspective of "stakeholders", meaning here individuals and communities that might be affected directly or indirectly by the company operations, products, services, and/or business relationships. The identification was tailored to the company in two, related columns:
   a. *Human Rights*

      For each ERM category, we identified which internationally recognized human rights the company had a non-negligible risk of adversely impacting through its products, operations, or business relationships.
   b. *Company-Related Risk*

      For each human right, we identified the specific ways in which the company may adversely impact it—and which stakeholders would be affected—with reference to the company's operations and the reasonable human rights expectations of a business (as opposed to a government).

When combining the first two columns, at least one dimension of the right to health: environment was a salient and material risk—albeit of differing degrees—in each of the Value Chain Categories.

iii. **Potential Involvement:** A company's involvement with any individual stakeholder risk is critical to determining its potential responsibility under the Guiding Principles, the CS3D, and the LkSG.[39] The more closely the company is involved, the greater its responsibility to address the harm. We identified the ways in which the company may be involved in the specific risk with reference to the categories in the company's ERM: direct, associated with the company's own policies and operations; and indirect, associated with the company's business relationships. These broadly align with the Guiding Principles, with "cause and contribute" as "direct" (i.e., resulting from the company's own actions or decisions) and "directly linked" as "indirect" (i.e., resulting from the actions or decisions of independent entities in the company's value chain). For analytical precision, we divided "indirect" into upstream and downstream business relationships, as the company's leverage and ability to address risks will differ between these relationship types.

Several risks potentially implicated distinct levels of involvement; for each, we identified all potential ways in which the company may be involved. With right to health: environment, the risk at different stages of the value chain implicated different levels of involvement—for instance, procurement risks were largely indirect, exploration risks were largely direct, and distribution risks were a combination of direct and indirect.

iv. **Assessment of Risk Level:** We assessed risk level to determine which risks to prioritize for the company's human rights strategy and program. We considered each risk level at two stages, inherent and adjusted. Inherent risk is projected based on the nature and scope of the company's products, operations, and value chain (including anticipated governance). Adjusted risk accounts for the actual governance the company has in place to manage inherent risk. Adjusted risk is most relevant for the purposes of practical strategy, so that resources are used most efficiently.

To provide coherence and precision, we divided the risks by "Salience" and "Materiality", using consistent definitions. We assigned each risk an "Impact" and "Likelihood" score, to fall into one of the 25 ERM matrix boxes. The scores thus offer transparency regarding method for more effective prioritization and efficient response.

v. **Practical Recommendations:** The disaggregated analysis provided by the first four columns enabled us to tailor risk-management recommendations to the specific function, risk dimension, and relevant protocol for most efficient alignment with key standards. The results could then be parceled out to risk owners

[39] LkSG §3(2)(4) ("(2) The appropriate manner of acting in accordance with the due diligence obligations is determined according to [...] 4. the nature of the causal contribution of the enterprise to the risk to human rights or environment-related risk or to the violation of a human rights-related or environment-related obligation.").

based on their specific expertise and authority, with subsequent monitoring integrated without need for any additional human rights expertise or capacity building.

The result of this process was a detailed governance improvement plan to meet the requirements of the most rigorous corporate human rights regulation facing the company—notably the CS3D and LkSG—while navigating potential litigation and reputational risk in multiple jurisdictions. While the ERM-based analysis was deployed for every right potentially affected by the company, the involved process of definition was only necessary for a handful of material and controversial rights, particularly those that might be vague and misused by plaintiffs' counsel. The whole ensured that rights most relevant to the company and its stakeholders were woven into the fabric of corporate governance while also providing the greatest resilience to potential future scrutiny by regulators, courts, and investors. This blend of art and science—drawing from abstract ideals to build practical, risk-tailored corporate programs—is increasingly the order of the day for effective sustainability advice. And it is a role for which lawyers are uniquely suited, if only we can approach the problems with creativity and humility.

References

AAA and others v. Unilever plc and another company (2017) EWHC 371, (2017) All E.R. (D) 07 (Q.B.)

Aftab Y (2014) The intersection of law and corporate social responsibility: human rights strategy and litigation readiness for extractive-sector companies, Rocky Mt. Min Law Inst 60:19–11

Aftab Y, Mocle A (2019) Business and human rights as law: towards justiciability of rights. Involvement and Remedy 66

Albareda L (2008) Corporate responsibility, governance and accountability: from self-regulation to co-regulation. Corp Gov 8:430–432

An Act to enact the Fighting Against Forced Labour and Child Labour in Supply Chains Act and to amend the Customs Tariff, Bill S-211, 1st Sess., 44th Parl., 2022 (Can.)

Araya v. Nevsun Resources Ltd., (2017) B.C.J. №2318, 419 D.L.R. (4th) 631 (B.C.C.A.)

Barber v. Nestlé USA, Inc., 154 F. Supp. 3d 954, 2015 U.S. Dist. LEXIS 170608 (C.D. Cal. December 9, 2015)

California Transparency in Supply Chains Act, Cal. Civ. Code § 1714.43 (West 2019)

Chandler v. Cape plc., (2012) EWCA Civ 525, [2012] 3 All E.R. 640

Choc v. Hudbay Minerals Inc., (2013) O.J. №3375, 2013 ONSC 1414, 116 O.R. (3d) 674 (Ont. S.C.J.)

Commission Proposal for a Regulation of the European Parliament and of the Council on Prohibiting Products Made with Forced Labour on the Union Market, COM (2022) 453 final (Sept. 14, 2022)

Dana v. Hershey Co., 180 F. Supp. 3d 652, 2016 U.S. Dist. LEXIS 41594 (N.D. Cal. March 29, 2016)

Das v. George Weston Lt., (2017) O.J. №3542 (Ont. S.C.)

European Parliament and Council Directive on Corporate Sustainability Due Diligence 2024/1760, 2024 O.J.

Garcia v. Tahoe Resources Inc., (2017) B.C.J. №117, 2017 BCCA 39 (B.C.C.A.)

Hunt P (2003) Report of the Special Rapporteur on the right of everyone to the enjoyment of the highest attainable standard of physical and mental health E/CN.4/2003/58

Kasky v. Nike, Inc. 45 P.3d 243 (Cal. 2002), cert. denied 539 U.S. 654 (2003)

Kerr M, Janda R, Pitts C (2009) Corporate social responsibility: a legal analysis, LexisNexis Canada

Leiferkettensorgfaltspflichtengesetz [LkSG] [Act on Corporate Due Diligence Obligations in Supply Chains], July 16, 2021, Bundesgesetzblatt, Teil I [BGBL. I] at 2959 (Ger.)

Loi 2017-399 du 27 mars 2017 relative au devoir de vigilance des sociétés mères et des entreprises donneuses d'ordre [Law 2017-399 of Mar. 27, 2017 on the Duty of Vigilance of Parent Companies and Ordering Companies], Journal Officiel de la République Française [J.O.] [Official Gazette of France] Mar. 28, 2017

Lov om virksomheters åpenhet og arbeid med grunnleggende menneskerettigheter og anstendige arbeidsforhold (åpenhetsloven) [Act relating to enterprises' transparency and work on fundamental human rights and decent working conditions (Transparency Act)], July 1, 2022, Lovvedtak 176 (2020–2021) (Nor.)

Lungowe and others v. Vedanta Resources plc and others (2017) EWCA Civ 1528, (2017) All E.R. (D) 102

Milieudefensie et al. v. Royal Dutch Shell plc, No. C/09/571932 (Hague Dist. Ct. May 26, 2021)

Modern Slavery Act 2015, c. 30 (UK)

Modern Slavery Act 2018 (Cth) (Austl.)

Okpabi and others v. Royal Dutch Shell plc and another, (2018) EWCA Civ 191, (2018) Bus. L.R. 1022

Pati R (2005) Rights and their limits: the Constitution for Europe in international and comparative legal perspective. Berkely J Int Law 223

Regulation 2017/821, of the European Parliament and of the Council of 17 May 2017 Laying Down Supply Chain Due Diligence Obligations for Union Importers of Tin, Tantalum and Tungsten, Their ores, and Gold Originating from Conflict-Affected and High-Risk Areas, 2017 O.J. (L 130) 1

Ruggie J (2011) UN Special Representative, Guiding Principles on Business and Human Rights: Implementing the United Nations "Protect, Respect and Remedy" Framework (UN Doc. A/HRC/17/31) (adopted by the UN Human Rights Council on June 16, 2011)

Sherman JF (2012) Shift Project, The UN Guiding Principles for the Corporate Legal Advisor: Corporate Governance, Risk Management, and Professional Responsibility https://www.americanbar.org/content/dam/aba/administrative/human_rights/sherman_legal_advisors_paper.authcheckdam.pdf. Accessed 24 Sept 2024

Spence DB (2011) Corporate social responsibility in the oil and gas industry: the importance of reputational risk. Chic Kent Law Rev 86:72–73

Sud v. Costco Wholesale Corp., 2016 U.S. Dist. LEXIS 5524 (N.D. Cal. January 15, 2016)

U.N Convention on the Rights of the Child, adopted 20 November 1989, entered into force 2 September 1990, 1577 UNTS 3, art. 24

U.N. Comm. on Econ. Soc. and Cultural Rts., General Comment No. 14 (2000): The right to the highest attainable standard of health (article 12 of the International Covenant on Economic, Social and Cultural Rights), U.N. Doc. E/C.12/2000/4

U.N. Convention on the Elimination of All Forms of Discrimination against Women, adopted 18 December 1979, entered into force 3 September 1981, 1249 UNTS 13, art. 12

U.N. International Convention on the Elimination of All Forms of Racial Discrimination, adopted 21 December 1965, entered into force 4 January 1969, 660 UNTS 195, art. 5(e)(iv)

U.N. Off. High Comm'r for Hum. Rts. (2012) The Corporate Responsibility to Respect Human Rights: An Interpretive Guide U.N. Doc. HR/PUB/12/02

UN Guiding Principles Reporting Framework, Salient Human Rights Issues: https://www.ungpreporting.org/resources/salient-human-rights-issues

Uyghur Forced Labor Prevention Act, Pub. L. No. 117-78, 135 Stat. 1525 (2021)

Votaw D (1973) Genius becomes rare: a comment on the doctrine of social responsibility Pt. II. Calif Manag Rev 15(5):11

White Paper, World Econ. Forum & Boston Consulting Grp., (2020) Embracing the New Age of Materiality: Harnessing the Pace of Change in ESG https://www.weforum.org/whitepapers/embracing-the-new-age-of-materiality-harnessing-the-pace-of-change-in-esg
Wirth v. Mars Inc., 2016 U.S. Dist. LEXIS 14552 (C.D. Cal. February 5, 2016)
Zalik A (2015) Transparency, auditability, and the contradictions of CSR. In: Walker-Said C, Kelly JD (eds) Corporate social responsibility? Human rights in the new global economy

Yousuf Aftab is a partner at Steptoe's New York office, where he focuses on business and human rights and sustainability law. With a rare depth of practical experience, Yousuf is renowned for delivering strategic and innovative counsel to Fortune 100 companies, governments, and international organizations on sustainability issues. He is consistently recognized as a Global Market Leader in Business & Human Rights Law by Chambers & Partners (2021 to 2025), with clients noting that "he is meticulous, focused on details and very strategic," with an "unparalleled" ability to predict market trends.

Yousuf has been at the forefront of business and human rights since 2008, when he served as counsel to the UN Global Compact. He has since conducted on-the-ground investigations and high-profile human rights assessments across the world, designed comprehensive corporate human rights policies and governance protocols, and managed disputes and crises, including US Congressional hearings and OECD specific instances. His clients have included many of the world's most prominent companies, including The Coca-Cola Company, HP Inc., GE Vernova, GE Aerospace, Novartis, Bayer, Barrick Gold, Teck Resources, Chevron, Exxon, Shell, Philip Morris International, Nestle, Credit Suisse, Deutsche Bank, EQT, NEI Investments, and Sustainalytics. He has also advised a range of governments and international organizations, including the OECD, UNICEF, and the Consumer Goods Forum, among others.

His expertise extends to the full spectrum of business and human rights laws and standards, from the CS3D, CSRD, LkSG, Norwegian Transparency Act, and various modern slavery acts to the TVPRA, Tariff Act, and UFLPA. He also has deep, practical expertise in the UN Guiding Principles on Business and Human Rights, and the OECD Guidelines for Multinational Enterprises.

Before joining Steptoe, Yousuf founded and led a Chambers-ranked boutique sustainability law firm and a sustainability consulting firm. Among his many contributions to the field, he co-authored the book "Business and Human Rights as Law" (LexisNexis, 2019), a work cited by the Supreme Court of Canada. Yousuf's career includes experience at esteemed international law firms in New York and Toronto, where he focused on international commercial disputes. He also played a key role in advising Debevoise & Plimpton LLP on the design and launch of their Business Integrity Group and clerked at the Supreme Court of Canada.

Qualified in New York and Ontario, Yousuf brings a unique combination of legal, strategic, and operational insight to help clients navigate the full array of sustainability challenges practically and efficiently.

Advancing the Power and Water Sectors

Aaron Atcheson

Abstract

Over the last twenty-five years, the power sector in Canada has changed radically, with more change to come as the country continues to electrify additional industries and components of life. The water and wastewater sector has begun to change as well, and with a significant infrastructure deficit and high expectations from the population, serious advancements will continue to be required. Lawyers working with various levels of government, private sector developers, manufacturers and sources of finance have all been instrumental in moving these industries forward. By learning from other jurisdictions and industries, Canadian lawyers have been able to assist their clients in adapting development and financial models to the local market and have made a big difference in the proliferation of renewable energy developments across the country. Moving forward, incorporating new carbon release reduction tools such as carbon capture and storage and hydrogen, will be critical, as well as dealing with significant new water contamination concerns. Lawyers working in these sectors will need to be more cognizant of the rights of indigenous groups, the priorities of all affected communities, and the tools at their disposal to progress infrastructure development.

1 Introduction

Twenty-five years ago, Canada's power sector was dominated by hydroelectric resources developed over the past century, a handful of nuclear power generating stations, and coal and natural gas-fired facilities heavily concentrated in areas of the

A. Atcheson (✉)
Miller Thomson LLP, London, ON, Canada
e-mail: aatcheson@millerthomson.com

A. McLaughlin, C. Chell (eds.), *Building the Case: How Lawyers are Leading on Sustainability*, https://doi.org/10.1007/978-3-031-88455-9_8

country without adequate hydroelectric capacity. With increasing demands for power, a population expecting change to address climate impacts, and governments looking to diversify methods of infrastructure delivery, the power industry was ripe for change.

Although it is popular to lament the role of lawyers in our society, very little change is possible without them. Lawyers were critical in writing new policy, working through the details of new programs for the development of electricity generating infrastructure, phasing out higher carbon intensity methods of producing power, creating the regulatory frameworks for development, assisting infrastructure developers with engineering, procurement and construction contracts, and assisting on a wide range of infrastructure finance options.

Our water sector has started a modernization march as well, though at a much slower pace and with a more drastic infrastructure deficit. New players have joined the effort in modernizing this industry, but with a population wary of private ownership of water and water related infrastructure, and with many of the most difficult problems in small or remote communities, not all resources have yet been brought to bear to this challenge. As more tools are utilized, lawyers will be needed in greater numbers to be able to obtain maximum benefit from our efforts.

While it is true that there is a wealth of lawyers who spend their careers seeking to stop developments of various kinds, in the areas of power infrastructure and water and wastewater facilities, many more are invested in bringing advancements to our country and its population, while ensuring that this is done safely, with the interests of all communities in mind. With additional challenges arising, the next generation of legal leaders will need to continue these important efforts.

2 How It Started

2.1 My Background

Growing up in a close-knit family on a farm in Elgin County, south of London, Ontario, the members of my family were always invested in the interests and efforts of each other. When my younger sister agitated at our secondary school for recycling programs, we were all interested in how this would go. While there were bumps in the road, this was the beginning of an environmental movement at our high school. Aside from the beauty of our family property and my love of being in nature, the passion of my sister for environmental issues caused me to include environmental law in my course selections at McGill Law School, and in my articling year in Toronto.

Practising environmental law at a national firm in Toronto, focused on large transactions, pushed me toward the solicitor side of the practice, and I split my time between environmental and real estate matters. Of a class of 22, I was the only student interested in either area, so I was able to obtain great experience during articling, and into my associate years. I had excellent mentors on the environmental side; Larry Cobb, one of the top environmental lawyers in Canada, and our group head, Jim Harbell, a leading practitioner in each of environmental, municipal and energy law. While I had great mentors on the real estate side as well, Jim's leadership role in the provincial energy field was inspiring. After spending 2002–03 in Dublin practising project finance law with a focus on the energy sector, I returned to Toronto determined to become an expert in the renewable energy field. I attended my first Canadian Wind Energy Association (CanWEA) conference in Pincher Creek, Alberta in the fall of 2003 and my first Canadian Solar Industries Association conference in Toronto the next year; I was one of two lawyers at the first and the only at the second.

Aside from there being no lawyers attending at that time, these associations were filled with hard-working, optimistic people, who saw the challenge ahead, wanted to help, but were not afraid to make a living in the process. While I spent considerable time writing and speaking in the area, my time working with the Ontario caucus of CanWEA was rewarding and illuminating.

Over the last twenty years, I have had the opportunity to work with dozens of independent power producers on projects across Canada, and in other countries with the assistance of local counsel. I have also been able to work for a number of component suppliers, contractors, lenders, and other service providers. Having the opportunity to work on many sides of these projects has provided invaluable experience in understanding the most important issues from each perspective.

In the last few years, I have found an industry that has many similarities with the renewable energy industry of the early 2000s. In joining Miller Thomson LLP, I have been able to contribute to a leading environmental law practice with members across Canada, and an expertise in areas including drinking water, wastewater and water contamination. The water sector has many different facets, with important work across the board. Aside from projects treating drinking water and wastewater, and addressing contaminated water situations, there are many other active water subsectors, including water testing and industrial water treatment. All of these areas require finance, and of course, legal assistance.

2.2 New Sources of Power

Twenty-five years ago, power generation across Canada was dominated by hydroelectric power, nuclear generation and other thermal generation based on the burning of coal, oil and natural gas. While wind power has a long history of being used to power single-site machinery, in 2000 in Canada, there was very little wind connected to the power grid and even less solar. The largest national conferences for wind and solar in 2003 had only a couple hundred participants. But through programs in multiple provinces facilitating renewable energy in the coming years, a significant industry has been grown. Beyond wind and photovoltaic solar (solar panels), independent power producers and solution providers to large energy users have developed power projects based on a wide variety of technologies and resources, from biomass and biogas to geothermal to solar thermal and beyond. Energy storage, once an add-on to intermittent power sources, is now being added to grids without sufficient physical storage capacity (i.e. hydroelectric dams) to do more with less generation. Many of these sub-industries have become reasonably mature over the last twenty years.

2.3 Legal Approach to Challenges

No surprise to lawyers that work on projects regularly, new sources of power required significant legal work to bring these projects to commercial operation. Independent Power Producer (IPP) hired lawyers have spent considerable time winding their way through laws and regulations, especially in the environmental, municipal and energy regulatory areas, not intended to address various categories of new energy projects. Some updates and advances pressed by governments have had unintended consequences; for example, Ontario's *Green Energy Act*, which created an omnibus approval cutting through all of these areas, concentrated attention on its expedited approval appeal process and spawned an organized NIMBY (Not In My Backyard) movement which dragged every project through a costly appeal phase.

One of the challenges with early projects was financing. As there was no financing available on a traditional bank loan basis, lawyers were engaged in structuring joint ventures and limited partnerships to allow balance sheet-based financing amongst interested parties. Once the early projects were successful and long-term power purchase agreements (PPAs) from government parties were available, we were able to move to project financing, following the model used in the United Kingdom and internationally. This in turn spawned the development of what has been called the "Canadian model" in some other parts of the world, where pension funds and insurance companies have provided the project finance funding.

On the government side, the development of long-term (20 to 25 years in most cases, longer for hydroelectric projects) power purchase agreements was critical to the advancement of this industry. Without long-term, committed purchasers, on set, reasonable terms, and without good modelling of resources by consultants, lenders would never have been interested in the sector. It was critical to ensure that government power purchasers would execute documents directly with lenders, easing the process involved should a project fail and the lenders needed to step in. Simply put, if power was sold on a spot basis with projects as so-called "merchant" power plants (as does happen in the natural gas fired thermal generation sector in some jurisdictions), the industry would never have advanced seriously; renewable energy involves a high initial capital cost with low variable and maintenance costs, making stable long-term offtake arrangements critical. Similarly, if reasonable allocation of risks had not occurred, on issues such as who is responsible for lost revenue during system outages, etc., we might still be financing projects on the basis of the balance sheets of mega-corporations, and there would have been no widespread development of renewable energy projects.

2.4 Bumps in the Road

The fact that lawyers have been able to facilitate the advance of the renewable energy industry does not mean that there have been no bumps in the road. From the early days, the industry has been begging for a consistent approach, rules of the game being set for the long term. Unfortunately, in many Canadian jurisdictions, the opposite has been the case. Big booms and busts have occurred in many provinces, the most notable being the rise and fall of the Feed-In Tariff (FIT) program in Ontario (2006–2018) and the more recent project-killing connection rule changes in Alberta. The cancellation of Ontario's offshore wind program in February 2011, based ostensibly on environmental concerns but probably more related to election considerations, is an example of government creating a level of uncertainty that undermines an industry, with the related litigation going on for many years after. With the cancellation of the FIT program, in just one weekend, we had twelve clients lose seventy-one already contracted projects, from large wind to commercial solar to micro-hydro projects. While one can argue that some of these contracts were more lucrative than necessary, or that projects were approved without regard for maintaining prime agricultural land or other valid planning purposes, it is easy to understand that such dramatic moves can undermine the success of a sector. One bright spot on this issue has been Quebec, that through governments of several stripes, they have continued to issue RFPs that allow for continued success for developers—not to mention, supporting the wind industry-related manufacturing in the Gaspé region.[1]

This is not to say that all the issues have been caused by government. In fact, the "Wild West" mentality that seemed to pervade the industry repeatedly as jurisdictions developed their programs, was particularly troublesome and led to significant changes in programs and concerns about "fly by night" developers. In fact, it can be argued that some of the groups that locked up connection capacity in some jurisdictions, only to sell off those rights to project owners who were advancing more quickly, were the reason behind second generation PPAs including a myriad of rules restricting changes of ownership of projects between PPA award and construction. Some of the "cowboys" of the industry, who locked up more sites than they could possibly develop, construct and finance, helped to turn landowner communities against wind and solar.

One episode stands out in terms of serious bumps in the road, and that involved Canada being taken to the World Trade Organization (WTO) in respect of the domestic content requirements of Ontario's FIT program. This put the Stephen Harper Conservative government in a position of defending a program from the McGuinty Liberal government in Ontario, where many small "c" conservatives were critical of the resources spent on the renewable energy industry. While a full effort was made to defend the program, Canada lost the fight and Ontario's program was found to have violated international law.[2] In the context of other incentives in

[1] Global Business Reports (2014); Renewables Now (2024).

[2] World Trade Organization (2014).

other jurisdictions, the Ontario program was hard to justify. And in the end, this was another reason for manufacturers to delay decisions on setting up plants in Ontario. Once the aggressive domestic content requirements were wound up, many manufacturing and assembly operations were shuttered. Ontario would have been wise to look East to Quebec to see how their program for wind component manufacturing has been so successful and has not resulted in a similar international battle.

3 How It's Going

3.1 Boom and Bust, Lawyers Needed

Practicing in the area of any government program supported industry is, by its nature, a series of boom and bust cycles. It is rare for a jurisdiction to provide long term assistance when political parties do not agree with the methods to be used to incentivize new power generation, or new water or wastewater treatment. In the early days of the renewable energy sector in Canada, there were developers of all sizes. When the Ford government came to power in Ontario and ended the Feed-In Tariff program, as well as terminating any project that had not yet started construction, a number of groups left the jurisdiction or wrapped up their operations completely (Kaiser 2018). Many groups headed to Alberta and its lucrative private PPA market, while others moved into energy storage, both for private and publicly procured projects. But either way, there is no question that the number of independent power producers (IPPs) has shrunk distinctly since the cancellation of the Ontario program.

Coincident with this hollowing out of the IPP ranks, the number of lawyers active in this space has dropped dramatically as well. A number of renewable energy specialists that left their firms for more lucrative specialized practices ended up having to seek alternative employment, while others moved on to other sectors. Some of us were fortunate enough to withstand the storm by diversifying our practices and following clients to other jurisdictions. At this point, there are healthy programs for renewable energy procurement in a number of jurisdictions and reduced red tape in others, though it seems there will always be some level of government-originated upset in this industry.[3]

As it is clear there is no one solution to climate change, we should be pursuing power production of all sorts as well as considering how hydrogen/ammonia can contribute to our carbon release reduction goals. Heavy industry located in Central Canada will require the continued use of hydrocarbons for the foreseeable future (for example, steel manufacturers), so we should be supportive of the interim use of carbon capture and storage while these industries continue to modernize. It also appears to be time to add small modular nuclear reactors (SMRs) to our electrical supply system in Ontario. I have been fortunate to work with the Nuclear Waste Management Organization on its deep geological repository project, which

[3] Wang and Noel (2023).

addresses valid concerns about the long-term storage of waste nuclear fuel. All these methods of reducing carbon release will be necessary to reduce the impact of climate change on our environment, and to reduce the other effects of releases into the atmosphere, such as particulate emissions leading to increased asthma-related deaths.

On the water side, we continue to have a great need for new, expanded or replaced wastewater and drinking water treatment systems at First Nation communities across the country, and especially in Canada's North. While we have had governments who have made pledges to eliminate drinking water advisories and boil water orders, and while some advancement has been made in these respects, there remains much work to be done. Similarly, while capital expenditures have taken place to better the infrastructure available on the water side, it is only more recently that investments have been made in training and operating capacity for smaller communities and in particular those in remote areas. Training local operators and expanding supports for these critical workers must be one of our priorities in the coming decades.

3.2 What Was Accomplished

The following highlights some of the dramatic changes in the power industry landscape over the last twenty years:

- As of 2022, Ontario had 5575 MW of wind and 2669 MW of solar installed on its grid.[4]
- Current Ontario generation capacity according to Ontario's Independent Electricity System Operator (IESO) in Fig. 1
- Based on recent data, Canada ranked 8th in the world for installed wind energy capacity and 22nd in the world for installed solar energy capacity.
- Canada has 337 wind energy projects and 206 major solar energy projects across Canada, accounting collectively for 7.24% of Canada's electricity demand in 2022.
- Canada now has an installed capacity of 21.9 GigaWatts of wind energy, solar energy and energy storage installed capacity.[5]
- Generation from wind and solar sources represented 7.2% of Canada's total electricity in 2023, which is a 37.4% increase from 2016.[6]
- Ontario has gone from approximately 1% of electricity produced coming from "alternative power sources" in 2000 to 12.2% in 2023.[7]

[4] Government of Canada (2024).

[5] Canadian Renewable Energy Association (2024).

[6] Statistics Canada (2024a).

[7] Statistics Canada (2002, 2024b).

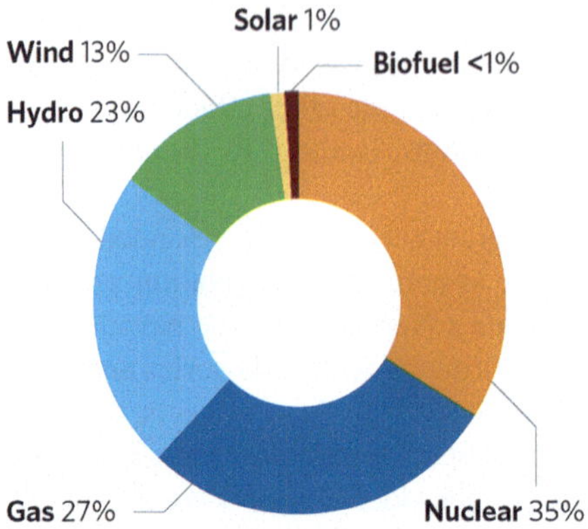

Nuclear	13,214 MW or 35%
Gas/Oil	10,471 MW or 27%
Hydro	8,862 MW or 23%
Wind	4,943 MW or 13%
Solar	478 MW or 1%
Biofuel	296 MW or <1%

Fig. 1 Current Ontario generation capacity according to Ontario's Independent Electricity System Operator (IESO) (Source: https://www.ieso.ca/power-data/supply-overview/transmission-connected-generation. Copyright © 2004–2022 Independent Electricity System Operator, all rights reserved). *Note that these figures do not include generators that operate within local distribution service, except for those that participate in the IESO-administered market. Most solar facilities in Ontario are currently connected to the distribution system (Independent Electricity System Operator 2024)

4 Where We Are Headed

4.1 Introduction

As many have remarked, change of many kinds continues to speed up. As climate change continues to be recognized as one of the world's greatest concerns, efforts to reduce carbon in the power industry will not abate. Lawyers are not simply assisting with the development of wind and solar projects, but are moving on to the next challenges.

In the mid-2020s the hot topic of the day is how artificial intelligence (AI) will move things forward, and how we put up appropriate safeguards to ensure that we do not destroy the foundations of our society and the infrastructure that has been built up over hundreds of years. In the areas of power and water, the stakes are incredibly high. Certainly we expect that AI programs will anticipate needs and allow us to do more with less resources, creating less waste and obtaining better value from our infrastructure. No longer should humans be responsible to sort through data for anomalies, as AI is deployed. No surprises like the old story about British power producers having to increase generation in advance of the end of Coronation Street, as every home put on a kettle.

While moving forward, it is critical that we not undermine important parts of the economy. As change continues to accelerate, policy writers and all responsible Canadians need to ensure that we safeguard necessary infrastructure from unrestricted change.

4.2 Carbon Capture and Storage

One of the areas where carbon release reductions threaten existing strategic industry is in the area of heavy manufacturing, such as steel production, and also in the areas of plastics, cement, and fertilizers. Decreases in Canada's aggregate carbon release through the drastic reduction of production from such industries would be a national disaster, and with competitive products coming from states not having the same carbon reduction goals or rules, unlikely to actually make a difference in terms of global climate warming.

Carbon capture and storage (CCS) is one of the powerful tools that could allow for significant decarbonization while protecting us from de-industrialization, especially where carbon is captured at the point of release. While carbon capture from the atmosphere has been advancing, the costs associated with capturing a tonne of carbon from early projects dwarfs the cost of capturing carbon at the plants where the carbon would otherwise be released.[8] At this point, we have decades worth of real world data on what can be captured, how it can be safely stored, and the associated costs.[9] We need to move past the broad statements of politicians to create the necessary regulatory environment for CCS projects, enable sources of funding and let the market pursue the best projects. One of the largest issues is the question of the ownership of "pore space", being the room that would be used to store the carbon deep below the surface, in appropriate geological formations. While provinces such as Alberta have declared that pore space is not owned by surface owners but is an asset of the province, in Ontario, where much of the country's heavy industry is located, the development of a CCS project would currently require the cooperation of hundreds or thousands of landowners, who own property in the area of the planned storage reservoir.

One of the risks in the CCS sector is that, while collection of carbon may happen in Canada, through failure to create the necessary conditions for storage projects in this country, the storage will actually happen in the rust belt of the United States. It will be up to the regulators, and certainly the lawyers involved in these efforts, to ensure that safe and long-term storage is available within reasonable proximity of Canadian industry.

4.3 Hydrogen and Ammonia

Hydrogen is at an interesting point in its evolution as a fuel and as a solution to certain of our power needs. While hydrogen fuel cells already power buses around the world, with Canada having one of the early players in this space with Ballard Power Systems Inc., hydrogen is spoken of as a fuel of the future. The media seems to want to point to hydrogen as the solution to all our power and carbon problems. Our politicians love to announce the possibility of new hydrogen projects and even

[8] Williams (2022).

[9] Cameron et al. (2023).

deals to provide hydrogen to other countries.[10] In reality, generating hydrogen is not inexpensive, and storing and transporting it is very difficult and expensive. Hydrogen should be viewed as a method of facilitating the energy transition, as opposed to a panacea for our carbon challenges.

Due to some of the qualities of hydrogen (i.e. H_2), most experts acknowledge that conversion to ammonia for storage and transport is necessary.[11] This conversion (and conversion back for fuel usage), plus the challenges of a fuel that holds less energy than natural gas or petroleum products mean that it is unlikely that all our energy needs will be addressed by hydrogen by 2050.[12] While it is true that a massive effort would need to be undertaken to create the infrastructure to allow for a significant component of our personal vehicles to be powered by hydrogen (as has been seen with electric vehicles (EVs) and EV charging infrastructure), that is not the only challenge. For commercial use, due to the costs of storage and transportation, the generation of hydrogen (or ammonia) is most likely to be successful if it is proximate to the industrial market for the fuel. If more hydrogen actually means "green hydrogen" (hydrogen created through zero emission sources of power), this means more green power production in the heavily industrialized parts of the country, such as southern Ontario. While so-called "blue hydrogen" (created through the use of natural gas) is often derided by environmentalists, blue hydrogen could be used through a transition period as we understand more about what we can and cannot do with hydrogen as a fuel.

For lawyers, any transition of infrastructure like this will require a significant effort to bring together parties interested in serving a new market (i.e. hydrogen fuel vehicle owners, hydrogen consuming industrial facilities), while minimizing the other changes that corporations and consumers will need to incorporate into their developments and lives. Hydrogen and ammonia hold great promise as part of the solution going forward, but it will require a significant amount of work both in the field and at boardroom tables.

4.4 Rediscovered Sources of Energy

As the old adage says, "everything old is new again". Never has this been more true than in the energy transition. The most obvious example is the renaissance of nuclear energy, with projects being considered across the country; the Canadian Nuclear Association reports that nine of ten provinces are currently considering nuclear power at some level, with Ontario and New Brunswick having operational reactors, and Alberta and Saskatchewan officially evaluating nuclear options.[13] Small modular reactors (SMRs) are one of the more popular recent topics of discussion, as Ontario has approved its first SMR at an existing nuclear generating station with

[10] Moore (2022).

[11] Serpell et al. (2023).

[12] Lan and Tao (2014).

[13] Canadian Nuclear Association (n.d.) Provincial Profiles https://cna.ca/provincial-profiles/.

more to come[14] and provinces who have previously ignored nuclear possibilities are now discussing projects.[15] With responsible long-term storage of spent nuclear fuel becoming more realistic, through the efforts of the Nuclear Waste Management Organization (NWMO) and their projects in Ontario, nuclear is poised to be a more acceptable source of power for this next generation.[16] Lawyers across the country will be boning up on the regulatory environment for nuclear, especially in relation to safe transport of nuclear fuel and waste, while others will be involved in massive construction and renovation projects allowing for more nuclear power to be added to the grid.

Another form of power that has gone through many media cycles as the "next big thing" may finally be to the point of providing additional green power to users. While the Bay of Fundy, with its world-leading tides, has been the proving ground for many first commercial scale tidal energy projects, and also their burial ground, it is possible that we may move beyond the hype in the coming decades. It is true that there is a massive source of untapped energy available in our oceans, and there should be ways to take advantage of this without negatively affecting the ocean ecosystem. As tidal projects become more prevalent, legal issues surrounding them will become more prominent. The balancing of competing rights, especially amongst energy producers and the fishing industries, both indigenous and otherwise, will come to the fore.

4.5 Tax Incentives for a More Mature Renewable Energy Industry

After over a decade of lobbying governments for a tax incentive plan in Canada similar to the successful one in the US, the federal government has been meting out the details of an investment tax credit program here. Whether this will create another influx of capital to the industry remains to be seen, but the initial interest is promising. Tax incentives are historically less likely to be used for political theatre than the awarding of contracts at elevated construction-incentivizing prices, and as time goes on, this type of long term support is understandable to all of the small, medium and large players in an industry, and to taxpayers. Unfortunately, some of the details of the new program have been slow to be clarified, and our federal government seems bent on adding an extra level of interference through the process of deeming which methods of energy production are appropriate, even with proven impacts in carbon reduction. Hopefully, future governments will invest the time and resources to mature this program, strip out some of the ideological elements, and focus on inducing private sector participants to take on the kinds of risks necessary for successful projects, where governments often languish.

[14] Government of Ontario (2023).

[15] Cole (2024).

[16] Nuclear Waste Management Organization (n.d.) Canada's deep geological repository. https://www.nwmo.ca/Canadas-plan/Canadas-deep-geological-repository.

As to whether tax incentive-based financing can be part of the solution on the water side, this remains to be seen. As a necessity of life, the water infrastructure sector is more complicated; for example, Canadians do not appear ready to allow profit driven corporations to have control of water supplies, outside of a limited number of water bottling operations. However, a tax incentive-based program to induce contaminated groundwater clean-up projects could be what is needed to move certain sites forward, especially in cases where funds are difficult to obtain and a pay-off through the sale of post-remediation properties is uncertain.

4.6 Water Contamination

In respect of water, we have a potential tsunami heading towards us through the twin problems of microplastics and Perfluoroalkyl and Polyfluoroalkyl Substances (PFAS). These concerns could result in the bankruptcy of major corporations and re-centre some of our views on what it is to be healthy and maintain a healthy population. Unfortunately, we will be dealing with the legacies of these contaminants for decades and even centuries. It will be critically important for us to address these problems as soon as possible. A study conducted by the International Journal of Environmental and Public Health found PFAS in the blood of 97 percent of Americans tested.[17] When you combine the pervasiveness of the exposure with the difficulty in dealing with these substances, and consider the ailments associated with exposure, this is clearly a very serious issue with no easy fix.

Similarly, microplastics have been in the news regularly, identified as accumulating in humans through fish and other foods[18] and contributing to the plaque in blood vessels of humans.[19] Serious changes are needed on this issue as well. While governments consider next steps and legal claims work through the courts, manufacturers continue to contribute to these problems. Faster action is required to limit the magnitude of these potentially massive issues.

4.7 Legal Personhood for Waterways

One of the issues that I believe will become an important part of project development over the coming years is the granting of legal personhood to elements of nature, and how this plays into the appeal of project permits. Largely through a failure of our systems to contemplate non-Western concepts of interest in our project consideration and approval processes, around the world indigenous groups and other activists have been pressing for the recognition of legal personhood for elements of nature. The most advanced of these is the press for the recognition of legal personhood for rivers and other waterways. In February 2021, the local Innu of

[17] National Institute of Environmental Health Sciences (2024).

[18] Alberghini et al. (2023), p. 789.

[19] Marfella et al. (2024), pp. 900–910.

Ekuanitshit and the regional municipality in Côte-Nord, Québec, passed a resolution granting legal personhood to the Magpie River (also known as the Muteshekau Shipu).[20] More recently, a suit has been filed in the Manitoba Court of King's Bench seeking recognition of Lake Winnipeg as a legal person, in order to force a comprehensive environmental assessment in respect of industry in the area of the lake.[21]

My principal concern is that legal personhood will be utilized at a late stage in the development process to force an additional appeal of a key project approval, based on the failure of the approval authority to take into consideration the needs of the waterway itself. I am also concerned that the individuals granted the right to represent elements of nature are motivated by concerns other than simply delaying or stopping a particular development. This movement should cause our governments to consider how they can take this perspective into account early in the planning and consideration process. It would be a failure for our environmental assessment and approval authorities if this movement results in an additional multi-year appeal at the end of every major project approval process. I had the opportunity to be included as a speaker at the House of Commons Standing Committee on Environment and Sustainable Development in a hearing on freshwater issues on January 30, 2024, where multiple viewpoints were considered on this issue. I expressed a strong preference to have governments engage on legal personhood for water and to find ways to incorporate the perspective of the waterways early in the approval process. I believe the federal government should also provide guidance in creating an appropriate trustee appointment process such that the group of individuals who speak for a waterway include those who are seeking solutions for problems, and not simply those looking to cancel a project being built in their backyard.

4.8 Global Water Scarcity

Global water security is an issue that will undoubtedly feature strongly over the second half of this century. As temperatures rise and clean water becomes more scarce, there will be temptations to commoditize clean water at various levels of government and through private parties. But unlike most other commodities, drinkable water is essential for life and like no other commodity. While few wish for new constitutional discussions in Canada, adding clean water as a constitutional right for all citizens and residents of Canada, should be top of mind if we ever re-open Pandora's box. The right to clean drinking water and sanitation has been recognized by the United Nations General Assembly.[22] Upholding this principle and ensuring that our international relations are consistent with it, will be more and more challenging in the coming decades.

[20] McAleer A (2021) Quebec's Magpie River is Now a Legal Person – A Monumental Moment in Canadian Environmental Law. https://cela.ca/blog-quebecs-magpie-river-is-now-a-legal-person/.

[21] Mary Maytwayashing et al. v. The Government of Hydro and Manitoba Hydro, Manitoba Court of King's Bench, File CI-24-01.

[22] Resolution 64/292, July 28, 2010.

Blessed with an abundance of clean water, Canada is also the home of a number of significant players in the water industry. While our diplomats have an important role in ensuring that water does not become a tool of war, there are also many opportunities to expand the role of Canadian water treatment technologies in providing safe and clean drinking water to communities around the world.

5 Conclusion

As a facilitator in one of the most active areas of change in the last twenty years, being a lawyer for independent power producers and lenders to the renewable energy industry has been challenging and rewarding, and I know that many of my colleagues feel similarly. It has been a pleasure to work with groups that have identified a problem and want to be part of a solution, but recognize that it is acceptable to make a profit on their successful efforts. The water industry is another sector that needs an appropriately guided industry effort, with significant challenges related to our severe infrastructure gap and many sources of serious contamination. Without lawyers and their efforts on behalf of clients, and in support of a more advanced society, little would have been accomplished since the turn of the century. These industries will continue to have exciting advancements over the next several decades, providing interesting opportunities for those lawyers engaged on meeting these challenges.

References

Alberghini A, Truant A, Santonicola S et al (2023) Microplastics in fish and fishery products and risks for human health: a review. Int J Environ Res Public Health 20(1):789

Cameron L, Carter A, Sievert K (2023) Why the cost of carbon capture and storage remains persistently high. https://www.iisd.org/articles/deep-dive/why-carbon-capture-storage-cost-remains-high#:~:text=Carbon%20capture%20and%20storage%20(CCS,tCO2%20for%20diluted%20gas%20streams

Canadian Renewable Energy Association (2024) By the numbers. https://renewablesassociation.ca/by-the-numbers/

Cole A (2024) The nuclear option: N.L. needs more energy, and some advocates say nuclear power is a solution. https://www.cbc.ca/news/canada/newfoundland-labrador/nuclear-energy-pub-1.7321285

Global Business Reports (2014) Power Québec. https://www.gbreports.com/wp-content/uploads/2014/11/QuebecPower2014_IE.pdf

Government of Canada (2024) Ontario: clean electricity snapshot. https://www.canada.ca/en/services/environment/weather/climatechange/climate-action/powering-future-clean-energy/overview-ontario.html

Government of Ontario (2023) Ontario building more small modular reactors to power province's growth. https://news.ontario.ca/en/release/1003248/ontario-building-more-small-modular-reactors-to-power-provinces-growth

Independent Electricity System Operator (2024) Transmission-connected generation. https://www.ieso.ca/power-data/supply-overview/transmission-connected-generation

Kaiser GE (2018) Ontario cancels wind and solar contracts. Energy Regul Q 6(3)

Lan R, Tao S (2014) Ammonia as a suitable fuel for fuel cells. Front Energy Res 2
Marfella R, Prattichizzo F, Sardu C et al (2024) N Engl J Med 390(10):900–910
Moore M (2022) 'Hydrogen alliance' formed as Canada, Germany sign agreement on exports. https://www.cbc.ca/news/canada/newfoundland-labrador/canada-germany-hydrogen-partnership-nl-1.6559787
National Institute of Environmental Health Sciences (2024) Perfluoroalkyl and Polyfluoroalkyl Substances (PFAS). https://www.niehs.nih.gov/health/topics/agents/pfc
Renewables Now (2024) Wind projects totalling 1.55 GW succeed in Hydro-Quebec tender. https://renewablesnow.com/news/wind-projects-totalling-155-gw-succeed-in-hydro-quebec-tender-846802/
Serpell O, Hsaln Z, Chu WY et al (2023) Ammonia's role in a net-zero hydrogen economy. https://kleinmanenergy.upenn.edu/research/publications/ammonias-role-in-a-net-zero-hydrogen-economy/#:~:text=Hydrogen%20can%20be%20stored%20in,heating%20ammonia%20to%20high%20temperatures
Statistics Canada (2002) Ontario's System-Wide Electricity Supply Mix: 2000 Data. https://www.oeb.ca/documents/electricity_mix.pdf
Statistics Canada (2024a) Hydroelectricity generation dries up amid low precipitation and record high temperatures: electricity year in review 2023. https://www.statcan.gc.ca/o1/en/plus/5776-hydroelectricity-generation-dries-amid-low-precipitation-and-record-high-temperatures
Statistics Canada (2024b) Ontario's System-Wide Electricity Supply Mix: 2023 Data. https://www.oeb.ca/sites/default/files/2023-supply-mix-data-update.pdf
Wang J, Noel W (2023) Investment impact of Alberta's Renewable Energy Moratorium: Renewable Energy Projects Fact Sheet. Pembina Institute. https://www.pembina.org/pub/investment-impact-albertas-renewable-energy-moratorium
Williams E (2022) The economics of direct air carbon capture and storage. https://www.globalccsinstitute.com/wp-content/uploads/2022/07/Economics-of-DAC_FINAL.pdf
World Trade Organization (2014) Canada – measures relating to the feed-in tariff program. https://www.wto.org/english/tratop_e/dispu_e/cases_e/ds426_e.htm

Aaron Atcheson is a partner in the London, Ontario office of Miller Thomson LLP, the leader of the firm's Projects Group, and the national leader of Real Property and Development. Aaron practices in the area of real estate and business transactions, and energy and infrastructure projects, with a focus on the power and water sectors.

Aaron started his legal career with Stikeman Elliott LLP in Toronto, practising real estate and environmental law with a focus on the energy sector. He spent one year with Matheson LLP of Dublin, Ireland (then Matheson Ormsby Prentice LLP) practising project finance law. He joined Miller Thomson LLP in 2007. Aside from leading the firm's cleantech, project development and real estate teams, Aaron has been Managing Partner of Miller Thomson's Southwestern Ontario offices, Chair of Miller Thomson's National Real Estate Practice Group, and a member of the firm's Executive Committee.

Aaron has been repeatedly recognized as a leading lawyer and expert in energy, environmental law and property development. He speaks and writes regularly in the areas of infrastructure development, including a regular legal corner in *Water Canada*, the leading national publication for that sector. Aaron and his wife Tiffany have raised three sons and Aaron runs, and coaches and plays basketball, in his free time.

Healing Tomorrow: Designing Sustainable Healthcare in a Complicated World

Jennifer D. Suess

Abstract

As someone with a lifelong passion to improve the patient experience of accessing timely and reliable healthcare, I have always been keenly interested in understanding the differing approaches used by government and private sectors around the world in establishing, and evolving, the necessary infrastructure to support this critical human need. As we now face global challenges including growing and aging populations with more complex medical needs than ever before, and seek to apply lessons learned during healthcare crises such as the devastating COVID-19 pandemic, there has never been a more essential time for lawyers around the world to contribute their unique combination of understanding legal regimes, coupled with strong advocacy skills, to lead the changes we all need to ensure a healthy future. Understanding how aged and inadequate healthcare models no longer meet the needs of the people they serve, as well as the complex and costly medical realities of increased life expectancy, insufficient human healthcare resources, and delayed implementation of technological advances, increasing access and creating more consistent and equitable outcomes for patients worldwide is an emerging area of focus for lawyers looking to make a difference in their careers. By practicing in a variety of corporate, not-for-profit and healthcare settings, I have seen firsthand the positive differences that one can make by bringing a sense of urgency in seeking new healthcare models to be supported by both private and public sectors. In exploring these issues in greater depth, this chapter identifies how different models of public private partnerships have and must continue to be implemented around the world to improve timely

With contributions from Evan Kirsh, Director, Legal, RioCan REIT.

J. D. Suess (✉)
RioCan REIT, Toronto, ON, Canada
e-mail: jsuess@riocan.com

A. McLaughlin, C. Chell (eds.), *Building the Case: How Lawyers are Leading on Sustainability*, https://doi.org/10.1007/978-3-031-88455-9_9

access to more effectively delivered healthcare on a global basis. We then shift to a comparative analysis of the Canadian and American healthcare systems to illustrate two distinct approaches to providing services and how the best elements of each can be leveraged to create optimal infrastructure. A deeper dive into Ontario, Canada's most populous province, is then presented as a case study to highlight what progressive changes have been implemented to date, and what further innovation is necessary to improve access to the delivery of sustainable healthcare. Throughout this chapter, by exploring the vast possibilities through the lens of an ESG framework, I highlight and encourage the various ways in which lawyers are empowered, and increasingly relied upon, to create, advocate for and oversee the execution of, public and private sector initiatives to improve access to healthcare and create more effective delivery models. Overall, this chapter aims to illustrate how lawyers who wish to lead in the continued evolution of healthcare must continue to champion the broad application of ESG principles within the legal system to create a healthier world for us all.

1 Introduction

As the global population continues to grow, there are increasingly greater challenges placed on already constrained healthcare systems. Although each country's challenges are unique in some ways, providing timely access to high quality healthcare is and should be a fundamental right for every person. Since 1946, the right to access healthcare has been internationally recognized as a fundamental human right regardless of age, gender, socio-economic status or ethnic background.[1] Yet, according to the World Bank and World Health Organization, half of the world's population lacks access to essential healthcare services.[2] Further, the 2023 Universal Health Coverage Global Monitoring Report identified that over the past 20 years, less than one-third of countries have improved health service coverage.[3] It is increasingly challenging to provide access to healthcare due to a confluence of factors including the increased rate of chronic disease, aging infrastructure, increased lifespan, political instability and a global shortage of qualified healthcare workers. Addressing these challenges requires collaboration from all parties including governments, corporations, international agencies, policy makers, industry professionals and, perhaps most impactfully, lawyers.

On an international scale, Canada has a well-regarded healthcare system that is decentralized, universal and publicly funded.[4] In a recent survey comparing ten high-income countries, however, Canada ranked last in its residents' ability to access primary heath care.[5] On an individual level, the Canadian population is also

[1] World Health Organization (2008), p. 1.

[2] World Health Organization (2017).

[3] World Health Organization (2023a, b).

[4] Allin et al. (2020).

[5] The Canadian Press (2024).

becoming increasingly dissatisfied. Significantly, the Canadian healthcare system faces daunting challenges due to a variety of reasons, including an aging population, the high cost of new technology, the legacy of generations of graduating or qualifying insufficient numbers of healthcare providers to meet the needs of a burgeoning population that, in recent years, has grown exponentially from immigration, and unacceptably long wait times to receive treatment.[6]

There is, however, cause for optimism as we boldly face the future in these uncertain times. Specifically, within the legal community and beyond, there is an increasing emphasis placed on strategic solutions that reflect environmental, social and governance ("**ESG**") considerations throughout various sectors. Leaders in the global economy, and especially lawyers, have the ability to create positive change by embracing and advocating for a more modern and innovative approach to healthcare systems which no longer meet the needs of their evolving populations. When I began my practice over twenty years ago, I chose to gain experience working at a national law firm on complex transactions, with a primary focus on the healthcare sector. Doing so allowed me to gain a deep understanding of the financial implications and motivations for parties to come together and create unique models to evolve Canada's healthcare system in parallel to its growing communities. It also allowed me to observe how a healthy population supports a healthy and thriving economy, which is critical for any nation. In the first ten years of my practice, I helped to establish various progressive models of healthcare across Canada which, while unquestionably costly at the outset, are now seen to have ultimately delivered exponentially higher value back to the communities in which they operate. As further detailed below, creating opportunities for earlier diagnosis, more timely treatment and offering greater access to remote communities in rural Canadian areas all resulted in individuals healing faster, recognizing symptoms more quickly and being more willing to seek treatment when necessary. In turn, individuals who can avail themselves of these opportunities are able to be productive and contribute to their communities and economies more effectively, and for longer periods of time. As Dr. Tedros Adhanom Ghebreyesus, Director-General of the World Health Organization states, "the fact that so many people cannot benefit from affordable, quality, essential health services not only puts their own health at risk, it also puts the stability of communities, societies and economies at risk".[7] Strong legal leadership helps to ensure that policies are drafted inclusively, basic human rights are advocated for and corporations operate in a socially responsible manner, irrespective of industry. Additionally, the unique perspective that lawyers bring in their comprehensive understanding of global legal and regulatory frameworks, combined with their ability to envision and advocate for meaningful change within local communities, provides lawyers the privileged position of both analyzing and proposing creative yet practical solutions to these complex problems in our changing world. Lawyers that take into account ESG considerations have the ability to have

[6] Government of Canada (2019).

[7] World Health Organization (2023a, b).

successful and meaningful careers, while leading positive change at both a macro-level, globally, and micro-level, in their communities.

The focus of this chapter is to highlight opportunities for improvement within global healthcare systems and how ESG principles can be utilized to generate greater access to healthcare and, consequently, better outcomes for all. First, we will identify positive outcomes achieved in healthcare systems internationally by exploring various models of public private partnerships ("**PPPs**"). Second, we will examine key similarities and differences between the Canadian and American healthcare systems. Third, we will describe how global healthcare issues are impacting Canadians, specifically in Ontario, and how the introduction of *Bill 60* incorporates elements of privatization within the existing socialized system and attempts to prioritize timely access to healthcare. Lastly, as Ontario licensed lawyers, we will use Ontario as a case study to provide recommendations focused on generating enhanced access to healthcare.

2 PPPs and How Different Models Can Create Positive Change

2.1 Understanding the Basics of PPPs

As one example of a model in which I have worked to advance in Canadian healthcare, PPPs are a form of long-term agreement between a government and private entity whereby the parties collectively contract to provide public services.[8] Although historically, PPPs are more commonplace in the transportation and energy sectors, governments around the world are becoming increasingly interested in PPPs to mobilize funds and promote reforms within healthcare systems.[9] PPP models often vary; however, they typically share a number of common features, including the use of private financing for capital and expenditures, the bundling of outputs and activities within the scope of a single contract between public and private entities, and the sharing of costs, risks and benefits between the contracting parties.[10] Bidding processes are typically used to find the best private partner based on an assessment of a shared vision, agreed upon resource allocation and cultural fit. Such an approach stimulates competition and encourages innovation and efficiencies to be identified amongst bidders, resulting in positive outcomes for the population being serviced.

There are three unique PPP models. Notwithstanding each model's differences, the implementation of PPPs in the delivery of healthcare services, including within hospitals around the world, has produced favourable outcomes with respect to cost-effectiveness and the performance of the specific healthcare unit.[11] Lawyers play pivotal roles within PPP projects at all stages—from structuring the initial

[8] Abuzaineh et al. (2018), p. 5.

[9] World Health Organization (2023a, b).

[10] World Health Organization Regional Office for Europe (2023), p. 9.

[11] Rodrigues (2023), p. 3.

transaction and drafting and negotiating agreements, to overseeing the execution of and facilitating the closing of the transaction. Substantively, PPPs also intersect a number of legal fields such as real estate, finance, securities and construction. Lawyers who are subject matter experts in these areas can work on complex and intricate transactions, while also generating positive and meaningful change within the jurisdictions in which they practice.

2.2 Model 1 PPPs: Discrete Clinical Service Model

The discrete clinical services model ("**Model 1 PPPs**") is a PPP whereby the government entity contracts with a private partner to operate and deliver specific services which are often performed at public healthcare facilities.[12] Model 1 PPPs are typically implemented in regions which experience limited capacity or access to a specific type of specialized service, such as diagnostics, radiology or dialysis.

Model 1 PPPs have been implemented successfully in multiple countries. In India, specifically at Sawai Man Singh Hospital in Jaipur, Model 1 PPPs are used for diagnostic services such as CT scans and MRIs which are not otherwise readily available in public facilities.[13] Similarly, the Cardiac Centre Facility in Deen Dayal Upadhyay Hospital in Dehradun was also created to enhance access for cardiac services.[14] In Romania, the government used Model 1 PPPs to award private contracts to four companies to operate and manage dialysis centres across eight hospitals.[15] Not only did the Romanian government increase access to and improve the quality of dialysis services, it also saved between €3–5 million between 2005 and 2008.[16] These examples demonstrate just a few of the myriad ways Model 1 PPPs can be innovatively applied to enhance healthcare infrastructure, leading to increased access and faster diagnosis and treatment, all of which ultimately leads to healthier communities.

2.3 Model 2 PPPs: Infrastructure Based Model

The infrastructure based model ("**Model 2 PPPs**") is a type of PPP whereby the private partner is contracted to build, rebuild or replace a public asset, and is tasked with maintaining the infrastructure throughout the lifespan of the contract.[17] Conversely, the government maintains responsibility to manage and provide clinical services.[18] Model 2 PPPs are typically implemented to enable access to private

[12] Abuzaineh et al. (2018), p. 26.

[13] Shah et al. (2018), p. 257.

[14] Shah et al. (2018), p. 257.

[15] Abuzaineh et al. (2018), p. 28.

[16] Abuzaineh et al. (2018), p. 28.

[17] Abuzaineh et al. (2018), p. 21.

[18] World Health Organization Regional Office for Europe (2023), p. 18.

funding to finance major infrastructure projects where there are restraints on public budgets. By building new infrastructure or modernizing existing infrastructure, more people gain access to quality health services.

Model 2 PPPs are commonplace in the United Kingdom to address infrastructure shortages.[19] For example, the British government launched a £1.1 billion hospital contract in 2006 for the redevelopment of St. Bartholomew's and The Royal London New Hospitals.[20] This was the largest private finance project in the United Kingdom and was completed in 2016.[21] It resulted in enhanced operational effectiveness, improved both general and specialist acute hospital services and provided both research and teaching opportunities.[22] This project also had a positive impact in the community by contributing to the regeneration of East London as a whole.[23] Model 2 PPPs have also been implemented in Canada with the development phase of a new hospital, Elder Care Lodge, and a new Ambulatory Care Centre on Moose Factory Island, having commenced more recently. This project is within the Weeneebayko Area Health Authority which provides healthcare services to people living in northern, remote regions of Ontario, specifically those surrounding the coast of James Bay and Hudson's Bay, many of whom are of Cree lineage.[24] Whether it is building new infrastructure in remote regions, or expanding existing aging facilities, Model 2 PPPs have been shown to be an effective and impactful vehicle for private capital to expand the capacity of, and provide additional healthcare services to, populations in need.

2.4 Model 3 PPPs: Integrated Model

The integrated model ("**Model 3 PPPs**") combines the core functions of both Model 1 PPPs and Model 2 PPPs, in that the private party both improves or expands healthcare infrastructure, but also provides certain healthcare services.[25] Given that Model 3 PPPs are more complex and less frequently used compared to other PPP models, they present a rewarding opportunity for lawyers to have a profound societal impact if they can be implemented correctly.

With the increase in PPPs globally, it is expected that Model 3 PPPs will also become more common. Its development can be attributed, in part, to situations like the following. After severe flooding left a substantial portion of the local population without adequate access to healthcare, one of the first Model 3 PPPs emerged in Valencia, Spain.[26] The private party not only financed and constructed La Ribera

[19] Ma et al. (2023), p. 472.

[20] UK Trade & Investment (2013), p. 4.

[21] UK Trade & Investment (2013), p. 4.

[22] European Investment Bank (2003).

[23] European Investment Bank (2003). https://www.eib.org/en/projects/pipelines/all/20030038.

[24] DCN-JOC News Services (2023).

[25] World Health Organization Regional Office for Europe (2023), p. 21.

[26] Abuzaineh et al. (2018), p. 31.

Hospital, but also, beginning in 1999, provided clinical services to the surrounding population of 250,000 residents.[27] A Model 3 PPP was also implemented in Maseru, Lesotho, with the opening of the Queen 'Mamohato Memorial Hospital in 2011. Although the publicly operated Queen Elizabeth II Hospital operated nearby, it was outdated and overcrowded. A study conducted by Boston University identified that the Queen 'Mamohato Memorial Hospital successfully treated a higher volume of inpatient and ambulatory patients, while averaging lower lengths of stay, as compared to Queen Elizabeth II Hospital.[28]

3 A Comparative Analysis of the Healthcare Systems in Canada and the United States

Growing up with both Canadian and American parents, lively debate often ensued around the dinner table as to which country's healthcare system provided optimal care for its citizens. Quick to recognize the irrefutable benefits of the universal healthcare system in Canada, my family often mused about how one might also import some of the advantages seen in the United States' Medicare and Medicaid systems without compromising the hallmarks of the Canadian regime. While acknowledging that many of the benefits seen in the U.S. are typically reserved only for Americans with private health insurance (and are therefore not equitably distributed), for these fortunate individuals, factors such as faster access to specialists, less waiting time in Emergency Rooms, higher availability of primary care physicians and the correlated faster diagnoses and treatment allowed by this approach are generally seen as advantageous in America. In becoming a lawyer, I have always searched for ways in which to evolve the Canadian healthcare system while keeping some of the above principles in mind. The ESG landscape now offers us a perfect platform from which to advocate for the continued adaptation of both systems to better serve the needs of their respective citizens, which can only be accomplished by having a deep understanding of what works (and what doesn't) in each of these countries.

3.1 A Detailed Overview

As noted above, the Canadian and American healthcare systems represent two distinct approaches to providing access to healthcare services, each with its own strengths and weaknesses. Canada has a decentralized, universal, publicly funded healthcare system called *Medicare*, created under the *Canada Health Act*.[29] As a result, more than 70% of healthcare spending is publicly funded through general tax

[27] Comendeiro-Maaloe et al. (2019), p. 408.

[28] Abuzaineh et al. (2018), p. 32.

[29] *Canada Health Act*, RSC 1985, c C-6.

revenues in Canada.[30] Pursuant to the *Constitution Act, 1867*, healthcare is primarily the responsibility of the provincial and territorial governments, which administer, and deliver healthcare services, and are responsible for the construction and maintenance of healthcare infrastructure.[31] Each province and territory has an insurance plan which theoretically covers all medically necessary hospital and physician services, whereas supplementary services (being those not covered under Medicare, such as prescription drugs, vision care and the services of other health professionals such as physiotherapists, chiropractors and occupational therapists) are largely financed by the patient out-of-pocket and/or through employer-based insurance policies of varying coverage.[32] The federal government plays a supporting role in the Canadian healthcare system. It provides certain funding, sets national standards through regulation (for example, for pharmaceuticals and medical devices), and is responsible for consumer safety and disease surveillance and prevention.

Most hospitals in Canada are publicly funded institutions with a fixed budget determined annually by the applicable provincial government. Healthcare providers typically operate on a fee-for-service payment system, whereby the provider bills the provincial or territorial government or private insurers, as applicable, directly for a predetermined fee based on the service performed.[33] The Canadian Medicare model aims to incentivize healthcare providers to deliver more services and seek higher patient volumes; however, frequent criticism includes the fact that it can lead to poor patient experiences.[34] Due to aging infrastructure, strict limits on human healthcare resources including the number of qualified physicians allowed to graduate annually, the restrictions on foreign trained doctors being able to requalify and practice in Canada in a timely and cost effective way, and an aging and growing population, healthcare providers and the system overall are overburdened, resulting in it becoming increasingly difficult for Canadians to receive timely, quality healthcare.

In a 2024 study, 65% of Canadians surveyed indicated that they could not receive same-day or next-day appointments when urgent care is required.[35] In addition, among a population of almost 40 million, nearly seven million Canadian adults do not currently have a family doctor nor other access to primary care.[36]As a result, among other issues, emergency rooms are overwhelmed, with the number of emergency room visits increased by 1.1 million from 2021–2022 to 2022–2023.[37]

By contrast, the United States does not have a universal healthcare system and is largely characterized by a multi-payer model, whereby healthcare services are

[30] Canadian Medical Association (2022a, b).

[31] Government of Canada (2019).

[32] Allin et al. (2020).

[33] Canadian Institute for Health Information (2020).

[34] Canadian Medical Association (2022a, b).

[35] Tasker (2024).

[36] Philpott (2024).

[37] Tasker (2024).

delivered by a combination of public and private entities.[38] There are two publicly funded American health programs—Medicare ("**U.S. Medicare**") and Medicaid. U.S. Medicare is a federal health insurance program for people 65 or older and also available to younger individuals with certain disabilities, regardless of their socio-economic status.[39] Alternatively, Medicaid is a federal and state program that helps cover medical costs for individuals with limited income and resources.[40]

Notwithstanding the existence of U.S. Medicare and Medicaid, the dominant form of coverage in the United States is private insurance, primarily offered through employers, which is particularly troubling, since this largely only covers portions of the population who are currently working or those who retire with extended health coverage.[41] Even with these various public and private programs, in 2023, 7.2% of Americans did not have insurance in any form, and by 2033, the number of uninsured Americans is expected to increase to include 8.4% of the population.[42] These metrics are further problematic since, as is seen elsewhere in the world, the cost of healthcare is increasing in the United States. Specifically, healthcare spending rose to $4.5 trillion in the United States in 2022, resulting in an average annual cost per person of $13,493.00 (assuming an approximate population of 333 million).[43] Moreover, this rise in cost has generally not been correlated with better health outcomes, which results in uneven quality of life across the American healthcare system, and is further exacerbated based on social determinants of health, such as income, gender and ethnicity.[44] For example, in 2020, the life expectancy of Hispanic and non-Hispanic Black groups declined more rapidly than non-Hispanic White groups.[45] Given the presence of multiple players, the healthcare system in the United States has become notably fragmented, with each player operating under its own set of rules, coverage criteria and reimbursement rates. The lack of a centrally coordinated and unified healthcare system results in varied healthcare delivery and quality, often deepening existing societal inequities.[46]

As shown above, the Canadian and American healthcare systems are more distinct than similar. While Canada's system is predominantly publicly funded, it generally results in longer wait times and fewer hospital beds per capita, leading to increased dissatisfaction among Canadians. Conversely, the United States' mostly private system costs much more to operate than the Canadian system, while leaving a substantial amount of its population uninsured, resulting in lower life expectancy and vastly uneven outcomes.[47]

[38] Donnelly et al. (2019), p. 1482.

[39] U.S. Department of Health and Human Services, Digital Communications Division (2022).

[40] U.S. Department of Health and Human Services, Digital Communications Division (2022).

[41] Peter G. Peterson Foundation (2023).

[42] Congressional Budget Office (2023).

[43] Peter G. Peterson Foundation (2024).

[44] Gunja et al. (2023).

[45] Agency for Healthcare Research and Quality (2022).

[46] Riley et al. (2012), p. 169.

[47] Ross University, School of Medicine (2021).

In the coming sections, we will highlight how Ontario, Canada's most populous province, is attempting to combat the issues facing its healthcare system and the critical role that lawyers must play in this much needed evolution. We will then highlight opportunities to generate further positive change.

4 The Move to Privatization in Ontario

4.1 Bill 60: Your Health Act, 2023

In an effort to improve the healthcare system, Ontario's *Bill 60, Your Health Act, 2023* ("**Bill 60**") received Royal Assent on May 18, 2023, and was proclaimed into force on September 25, 2023.[48] The preamble of Bill 60 explicitly provides that the Ontario population and government "intend to expand access to publicly funded community-based health services to improve patient wait times, patient experience and access to care in a way that considers the needs of diverse, vulnerable, priority and underserviced populations, taking into account linguistic needs."[49]

One key element of Bill 60 is the introduction of 'as-of-right' licensing rules which are created to address healthcare worker shortages by allowing regulated healthcare professionals (such as physicians and nurses) from other Canadian provinces and territories to temporarily practice in Ontario without the need of an additional approval or license.[50] This is just one attempt to increase the supply of healthcare professionals to catch up with Ontario's increased demand; however, the effects are not as expansive as initially anticipated.[51] First, 'as-of-right' licensing does not allow out of province healthcare professionals to provide certain services, like virtual care, to Ontarians. Rather, the services must be performed within an Ontario public hospital or long-term care home. Second, although it was discussed that 'as-of-right' licensing would circumvent the previous lengthy, complicated and costly process of allowing out of province healthcare professionals to practice in Ontario, it only does so temporarily for certain healthcare professionals who have already applied for their formal Ontario license. In an effort to further reduce barriers and increase access to healthcare, adopting a more expansive version of 'as of right' licensing would be a positive step towards a pan-Canadian medical license which, according to the Canadian Medical Association, is supported by 95% of physicians and medical students across Canada.[52]

An additional group of individuals who can support Ontario's inadequate healthcare infrastructure is pharmacists, who are widely considered to be highly skilled, trusted and regulated, but have not historically been leveraged as a potential resource

[48] Laviolette et al. (2023).

[49] *Your Health Act*, 2023, SO 2023, c 4.

[50] Reid et al. (2023).

[51] Laviolette et al. (2023).

[52] Canadian Medical Association (2023).

to alleviate some of the strain on Canada's medical system.[53] Bill 60's expanded scope of practice for pharmacists is another attempt by the provincial government to increase timely access to healthcare in Ontario. Specifically, Bill 60 revised *The Pharmacy Act, 1991* (the "**Pharmacy Act**"), to allow a pharmacist's practice to now include the assessment of conditions for the purposes of providing medication therapies and prescribing medication for 19 minor ailments including acne, nausea and vomiting in pregnancy and gastroesophageal reflux disease.[54] These changes are beneficial for Ontarians as pharmacists are often the most convenient resource for certain healthcare services. These changes are also intended to reduce wait times at more traditional healthcare facilities like walk-in clinics and hospitals in order to free up these facilities for more urgent matters. As of October 2023, 89% of Ontario pharmacists are participating in this new program, resulting in 400,000 pharmacist-led assessments being completed.[55] Although there is no requirement for pharmacists to expand their practice, certain major national corporations such as Shoppers Drug Mart, with pharmacies located across the Province of Ontario, have publicly applauded Bill 60's revisions to the Pharmacy Act as this change allows pharmacists to be a part of an innovative solution to Ontario's healthcare challenges.[56]

Another attempt to reduce wait times for healthcare services in Ontario is through the creation of 'integrated community health services centres'. With the passing of Bill 60, the *Integrated Community Health Services Centres Act, 2023* (the "**ICHSCA**"), will replace the *Independent Health Facilities Act* (the "**Old Act**").[57] Although the Old Act was in existence for approximately 30 years, no new licenses for private healthcare facilities were issued for nearly two decades.[58] The ICHSCA expands the definition of health facilities to include a broader range of services such as surgical or diagnostic centres, which were not captured under the Old Act. It is also expected that more licences will be issued by the provincial government under the ICHSCA, in particular for centres which offer these newly added services. Of particular note, the ICHSCA contains a new 'public interest' requirement that an applicant must satisfy prior to a licence being issued. Specifically, the applicant (who need not be a physician) must include an explanation as to how the proposed centre will provide connected and convenient care, including the applicant's capacity to improve patient wait times and distances to other centres and hospitals.[59] Certain critics of this approach note that this change pushes towards privatization instead of focusing efforts to improve the public system, and further worry that for-profit centres will overcharge or upsell customers, thereby siphoning provincial funding and healthcare professionals to the private sphere. Although these are valid concerns, processes can and should be implemented to protect Ontarians from these

[53] Canadian Pharmacists Association (2023).

[54] Ontario Health Coalition (2023).

[55] Government of Ontario (2023a, b).

[56] Shoppers Drug Mart (2023).

[57] Government of Ontario (2023a, b).

[58] Laviolette et al. (2023).

[59] Watts et al. (2023).

outcomes while continuing to focus on reducing wait times and improving access to health services.

Law students and lawyers who are well versed in these regulated areas have a unique opportunity to foster innovative change. For example, law students can participate in public awareness campaigns to educate community members on the effects of Bill 60. Moreover, lawyers can and should work closely with legislators to suggest necessary amendments to Bill 60 designed to amplify its positive impact, and to propose new impactful bills which aim to generate even further positive healthcare reform.

4.2 Privatization in Ontario

In addition to the introduction of Bill 60, other steps are also being undertaken by the provincial government aimed at creating better access to healthcare in Ontario, while trending towards privatization. For example, Ontario's 2022 budget contained a 'Plan to Build Ontario', identifying plans by the government to build and improve hospitals by investing more than $40 billion towards 50 major hospital projects (adding, among other resources, 3000 new beds over the next ten years).[60] The Plan to Build Ontario uses aspects of a Model 2 PPP, as described in Sect. 2.3 of this chapter, to allow a consortia of private companies to design, build, finance and maintain various sites across Ontario including Ottawa Hospital's new Civic campus, Mackenzie Vaughan Hospital and the addition to Toronto's SickKids hospital.

In an effort to meet the needs of its growing and aging population, the Ontario government is also focusing its efforts on increasing the supply of long-term care beds by committing to build 30,000 new long-term care beds by 2028.[61] By comparison, between 2011 and 2018, only 611 net new beds were built in Ontario.[62] As part of this initiative, the Ontario government is selling surplus provincial properties in the cities of Vaughan, Oakville and Aurora (which have a population of over 300,000, 200,000 and 60,000, respectively), on the condition that long-term care homes are built. The anticipated result of this project is that 896 new long-term beds will be available to Ontarians in need.[63] The Ontario government is also using innovative initiatives such as the Accelerated Build Program to deliver projects within Ontario's long-term care system on a condensed timeline of "months, not years" compared to the typical procurement process.[64] As Premier Ford stated, "our government is putting forward bold new solutions to turbo charge the development of long-term care beds across the province and ensure our most vulnerable citizens can live in modern surroundings."[65]

[60] Government of Ontario (2022a, b, c).

[61] Government of Ontario (2022a, b, c).

[62] Government of Ontario (2022a, b, c).

[63] Government of Ontario (2020a, b).

[64] Office of the Premier (2020).

[65] Government of Ontario (2020a, b).

These are clear efforts made by the Ontario government to improve access to healthcare for Ontarians. Although all initiatives and legislative changes are not without drawbacks, using innovative methods to address these issues is paramount.

5 Further Recommended Steps to Improve Access to Healthcare for all Canadians

Although some are justifiably hesitant to make certain changes to Canada's healthcare system, it is generally agreed upon that the existing framework does not meet the evolving needs of Canada's population. The methodology of evaluating the impact of any change should not be whether it is a departure from the existing system or structure but, rather, whether it will increase timely access to quality healthcare, while improving overall health outcomes as equitably as possible across the population. While the Ontario government has, and continues to, take positive steps to address certain systemic challenges, there is a critical opportunity for further enhancements to all facets of the healthcare system. As Canadian lawyers licensed in Ontario, this section will provide recommendations to improve Ontario's healthcare system, which can also be adapted and adopted in the U.S. and other jurisdictions. We encourage our readers to approach these challenges as advocates who can assist with this evolution in the coming decades.

To begin, in an effort to reduce healthcare expenditures in the future, Ontario should adopt proactive healthcare practices, based on preventative care, as opposed to reactive, treatment based care, as currently only 5% of Canadian health dollars are used for health "promotion and prevention", whereas the remaining 95% of Canadian health dollars are used for "sickness care".[66] The overarching goal of this philosophical change would be to reduce a person's likelihood of becoming ill and developing diseases which can, in part, be accomplished through expanded preventative services at the primary healthcare level, including more regular check-ups and disease screenings, in an effort to diagnose and treat illnesses when the likelihood of recovery is greater. In addition, more efforts should be made to focus on the implementation of healthy habits at a societal level which can happen in the partnership between a healthcare provider and their patients, empowering patients to make educated lifestyle decisions that prevent disease, extend lifespan and lead to fewer medical complications when facing illness. By implementing a broader range of holistic and preventative services, such as mental health and behavioural services within existing or new PPP projects, society will become healthier, and trust within the healthcare system will be improved, while alleviating the stress placed on the remainder of the existing infrastructure.

Expanding the availability of telehealth and virtual patient visits using technology would also drastically improve access to healthcare for many Canadians, especially those in remote areas of the country. The rates of telehealth use in Canada drastically increased from 10% to 20% of all healthcare visits in 2019, to 60% of all

[66] Picard (2024).

healthcare visits in Canada in 2020.[67] While increased utilization of this technology was initially borne out of necessity as a result of the COVID-19 pandemic, its proliferation should continue to be a focus of healthcare reform given the number of resulting positive outcomes. For example, telehealth can be used to increase access to healthcare for people living in rural and remote regions (such as in Canada's three territories which, as of 2021, represent 39.1% of Canada's landmass but only 0.3% of its population), who typically have access to a disproportionately low number of doctors and healthcare services.[68] Urban areas, as well, would benefit from enhanced telehealth services, that can be used for the delivery of diagnosis and treatment options, given the general lack of access to primary healthcare services for Canadians. Effectively implementing telehealth as a mainstay of healthcare systems globally would improve equality of care by improving access for portions of the population that are currently under-serviced. Telehealth also reduces barriers such as time and cost of traveling to appointments, encourages continuity of care for patients in need and reduces the number of emergency room visits where limited alternatives to healthcare exist.[69]

Although telehealth is just one method to modernize Canadian healthcare, the use of innovative technologies can benefit the healthcare sector in a number of significant ways. For example, the appropriate use of artificial intelligence ("**AI**") and machine learning already helps to automate the more administrative and data-oriented functions of the sector, which can alleviate time constraints of healthcare practitioners and allow them to assist more patients. In addition, innovation can be used to create systems that promote the centralization and sharing of data, resulting in more efficient patient experiences. In the future, it is entirely foreseeable that at-home diagnostic kits and enhanced AI will be leveraged to increase both the access to, and efficiency and quality of, the delivery of healthcare.

It is broadly and globally acknowledged that one of Canada's and Ontario's most attractive and growth-oriented features is its multicultural and diverse populations.[70] There are, however, inequitable gaps in access to healthcare by many immigrant populations, which result in overall health status inequalities. More efforts need to be made to foster inclusivity and true access for all. By embedding translation services, for example, as a mainstay throughout the entire healthcare system, individuals who do not speak English as a first language will feel more comfortable seeking help and will receive better treatment. Such services can be delivered virtually and at relatively low costs to the provider, especially when taking into account that better and more precise diagnosis leads to enhanced clinical outcomes, which ultimately saves the system money overall.

The location of traditional and new innovative healthcare services should also be carefully considered and embedded where Ontarians commonly live, work and shop. In the real estate industry specifically, owners and developers should actively

[67] Canada Health Infoway (2022).

[68] Statistics Canada (2022).

[69] Haleem et al. (2021).

[70] Statistics Canada (2021).

seek healthcare uses within their shopping centres or mixed-use developments to provide easy access to healthcare services for the surrounding populations. Whether it is a pharmacy, virtual healthcare kiosks or public or private clinics—resources must be allocated to implement healthcare uses into new and existing commercial and mixed-use spaces aimed at increasing timely access to healthcare services, and in turn, increasing the likelihood that these services will be utilized efficiently and effectively.

Another way to improve access to healthcare is through the creation of new healthcare services and the expansion of existing healthcare infrastructure. Whether it is privately or publicly funded and operated, the paramount concern should be increasing the supply of healthcare facilities and services to augment the surging demand. By privatizing certain healthcare functions, such as MRIs, CT scans and more routine surgeries such as endoscopies and joint replacements, wait times will be reduced and healthcare providers may choose to remain in Canada as their earnings, working hours and overall quality of care delivered are positively impacted. To be clear, the necessary inclusion of certain privatized services should not be intended to supersede the existing public system but, instead, to augment and work in tandem with the existing public system, while improving healthcare infrastructure for all.

Unique and expanded practices to service various patient populations should also occur within the existing public infrastructure. For example, expanding operating hours to evenings and weekends would help reduce wait times in many areas. This idea was already successfully implemented by Toronto Regional Arthroplasty Collaborative in an effort to reduce wait times, resulting in patients receiving hip and knee replacements in the region 25% faster.[71] Similarly, The Hospital for Sick Children in Toronto continues to leverage partnerships with community hospitals to utilize available surgical resources, as well as to run their operating rooms with extended hours, in an effort to address record-long waiting lists for children requiring surgery following the COVID-19 pandemic.[72]

Given the call for expanded healthcare facilities and practices, it is critical to also increase the mobility of existing healthcare service providers and the quantity of total healthcare service providers to keep up with demand. Increasing the mobility for Canadian healthcare providers would help address regional inequalities. For example, expanding Bill 60's 'as-of-right' licensing rules to further reduce the barriers for non-Ontario licensed healthcare providers to more easily and permanently provide healthcare services in a wider variety of healthcare facilities in Ontario (instead of limiting practitioners to a single or few sites in their provisions of service). Removing the requirement for physicians to be individually licensed in each province, and in turn implementing a pan-Canadian licence for healthcare providers would generate more favourable results Canada-wide. In addition, broadening jurisdictions of practice would also make Canada a more attractive place for internationally trained medical students and healthcare providers to work, particularly if regulations surrounding the transferability of credentials for foreign trained

[71] Ogilvie et al. (2024).

[72] Weeks (2024).

healthcare workers were relaxed. At the institutional level, expanding admissions to healthcare education programs would also allow the profession to, over the long term, better match the demand of a growing population. The positive effects of these changes would be further compounded if technology and data were creatively used to study long-term labour requirements and needs in the healthcare space. For example, this could be done by incentivizing medical students to pursue certain fields which are, or are expected to be, in high demand, like family doctors.

As depicted above, in order to ameliorate the many layered and complex facets of the healthcare sector, all functions and aspects should be continually and critically analyzed to understand and execute upon opportunities for improvement. Such an approach should include private insurance programs which are used to fill gaps for healthcare expenditures not covered under Medicare. Given the North American population, along with those of most developed countries, is aging, more focus should be placed to create innovative insurance coverages for seniors, who require a disproportionate amount of healthcare services and incur a disproportionate amount of healthcare costs. Specifically, in Canada, individuals aged 65 and older accounted for 45.7% of total healthcare expenditures in 2019, and given the aforementioned aging population, this statistic is expected to rise to 71.4% by 2040.[73] Particularly troubling is the accompanying reality that seniors who are no longer working earn less income, and therefore are often not able to afford healthcare services when in critical need of care. Similar to certain existing investment vehicles like Canada's registered retirement savings plan and the well-funded medical benefits programs enjoyed by many public sector workers in Canada, governments should motivate private insurance providers to create programs which are contributed to regularly during the course of one's employment, but are only accessible during retirement. Doing so would create a pool of funds to be drawn upon when additional care is required most in the later phases of life. For employers, partnering with these private insurance providers, and matching employee contributions, would greatly improve access to quality healthcare later in life and could result in increased, healthier longevity.

6 Conclusion

As we have only just begun to explore in this chapter, we stand globally as nations of people living longer, wholly focused on ways in which to improve the shared human experience of accessing and benefiting from efficiently managed, innovative and impactful healthcare. Although each country and region may face unique healthcare challenges, a healthy population is critically important to a thriving economy. For lawyers beginning their careers, the opportunity to leverage your skills of advocacy, problem solving and creativity have never been brighter. Using a platform anchored by the ESG principles of creating healthier societies, with efficiently managed and thoughtfully delivered healthcare, is one of the most rewarding and long-lasting contributions any lawyer can hope to make in one's career. Regardless of a

[73] Globerman (2021).

region's existing structure—PPP, public, private or otherwise, we encourage you to study globally relevant programs and policies which can be adopted to meet the needs of growing populations. Within Ontario, Bill 60 and other steps allowing for a hybrid approach involving a combination of publicly funded and privatized elements are useful tools to address a growing and aging population which is becoming increasingly dissatisfied with its historically renowned, but now out-dated, healthcare system. The recommendations provided in this chapter are obviously just some of many opportunities to improve access to healthcare, although the underlying principles can be adapted and implemented worldwide. We encourage current and aspiring lawyers to identify a sector they are passionate about, become subject matter experts within it, and actively advocate for the change necessary to create a healthier world.

References

Abuzaineh et al (2018) PPPs in healthcare: models, lessons and trends for the future. In: Healthcare public-private partnerships series, No. 4. PwC, The Global Health Group. https://www.pwc.com/gx/en/healthcare/assets/ppps-in-healthcare.pdf

Agency for Healthcare Research and Quality (2022) National Healthcare Quality and Disparities Report. https://www.ahrq.gov/sites/default/files/wysiwyg/research/findings/nhqrdr/2022qdr.pdf

Allin et al (2020) International Health Care System Profiles, Canada. https://www.commonwealthfund.org/international-health-policy-center/countries/canada

Canada Health Act, RSC 1985, c C-6

Canada Health Infoway (2022) Canadians' health care experiences during COVID-19: uptake of virtual care. https://www.infoway-inforoute.ca/en/component/edocman/3828-canadians-health-care-experiences-during-covid-19/view-document?Itemid=0

Canadian Institute for Health Information (2020) A profile of physicians in Canada, 2020. https://www.cihi.ca/en/a-profile-of-physicians-in-canada-2020

Canadian Medical Association (2022a) From fragmentation to integration: rethinking models of care. https://www.cma.ca/our-focus/fragmentation-integration-rethinking-models-care

Canadian Medical Association (2022b) Health care funding in Canada. https://www.cma.ca/latest-stories/health-care-funding-canada

Canadian Medical Association (2023) 95% of physicians, medical learners support pan-Canadian licensure to improve access to care: CMA poll. https://www.cma.ca/about-us/what-we-do/press-room/95-physicians-medical-learners-support-pan-canadian-licensure-improve-access-care-cma-poll

Canadian Pharmacists Association (2023) What pharmacists can do across Canada. https://www.pharmacists.ca/advocacy/scope-of-practice/

Comendeiro-Maaloe et al (2019) Public-private partnerships in the Spanish National Health System: the reversion of the Alzira model. Health Policy 123(4):408–411

Congressional Budget Office (2023) Federal Subsidies for Health Insurance: 2023 to 2033. https://www.cbo.gov/system/files/2023-09/59273-health-coverage.pdf

DCN-JOC News Services (2023) Development phase underway for WAHA redevelopment. https://canada.constructconnect.com/dcn/news/infrastructure/2023/01/development-phase-underway-for-waha-redevelopment

Donnelly et al (2019) Single-payer, multiple-payer, and state-based financing of health care: introduction to the special section. Am J Public Health 109(11):1482–1483

European Investment Bank (2003) Royal London and Barts Hospitals PPP project. https://www.eib.org/en/projects/pipelines/all/20030038

Globerman S (2021) Aging and expenditures on health care. https://www.fraserinstitute.org/sites/default/files/aging-and-expenditures-on-health-care.pdf

Government of Canada (2019) Canada's Health Care System. https://www.canada.ca/en/health-canada/services/health-care-system/reports-publications/health-care-system/canada.html#a9

Government of Ontario (2020a) Ontario accelerating the development of long-term care homes. https://news.ontario.ca/en/release/57613/ontario-accelerating-the-development-of-long-term-care-homes

Government of Ontario (2020b) Ontario selling surplus properties to build three new long-term care homes. https://news.ontario.ca/en/release/59047/ontario-selling-surplus-properties-to-build-three-new-long-term-care-homes

Government of Ontario (2022a) Ontario increasing construction funding for long-term care homes. https://news.ontario.ca/en/release/1002523/ontario-increasing-construction-funding-for-long-term-care-homes

Government of Ontario (2022b) Ontario on track to build 30,000 new long-term care beds. https://news.ontario.ca/en/release/1002083/ontario-on-track-to-build-30000-new-long-term-care-beds

Government of Ontario (2022c) Ontario's plan to build. https://news.ontario.ca/en/release/1002123/ontarios-plan-to-build

Government of Ontario (2023a) Even more convenient care close to home coming to communities across Ontario. https://news.ontario.ca/en/release/1003584/even-more-convenient-care-close-to-home-coming-to-communities-across-ontario

Government of Ontario (2023b) Ontario's Regulatory Register. https://www.ontariocanada.com/registry/view.do?postingId=44708&language=en

Gunja et al (2023) U.S. health care from a global perspective, 2022: accelerating spending, worsening outcomes. https://www.commonwealthfund.org/publications/issue-briefs/2023/jan/us-health-care-global-perspective-2022

Haleem et al (2021) Telemedicine for healthcare: capabilities, features, barriers, and applications. Sens Int 2:100117

Laviolette et al (2023) Part 4 –Your Health Act, 2023: what you need to know now that Bill 60 is in force. https://www.blg.com/en/insights/2023/10/your-health-act-2023-what-you-need-to-know-bill-60

Ma et al (2023) Are public—private partnerships still an answer for social infrastructure? A systematic literature review. SpringerLink 10:467–482

Office of the Premier (2020) Ontario unveils new innovative approach to building long-term care homes faster. https://news.ontario.ca/en/release/57690/ontario-unveils-new-innovative-approach-to-building-long-term-care-homes-faster

Ogilvie et al (2024) A new hip in 4 weeks instead of months: these Toronto doctors' plan to cut surgical wait times is seeing results. https://www.thestar.com/news/canada/a-new-hip-in-4-weeks-instead-of-months-these-toronto-doctors-plan-to-cut/article.html

Ontario Health Coalition (2023) Analysis: Ontario Health Coalition Analysis of Bill 60, Your Health Act. https://www.ontariohealthcoalition.ca/index.php/analysis-ontario-health-coalition-analysis-of-bill-60-your-health-act/

Peter G. Peterson Foundation (2023) The share of Americans without health insurance in 2022 matched a record low. https://www.pgpf.org/blog/2023/11/the-share-of-americans-without-health-insurance-in-2022-matched-a-record-low#

Peter G. Peterson Foundation (2024) Why are Americans paying more for healthcare? https://www.pgpf.org/blog/2024/01/why-are-americans-paying-more-for-healthcare

Philpott (2024) It's time for Canadians to have the right to a family doctor. https://www.theglobeandmail.com/opinion/article-its-time-for-canadians-to-have-the-right-to-a-family-doctor/

Picard A (2024) Some sensible advice on health spending, still ignored 50 years later. https://www.theglobeandmail.com/opinion/article-some-sensible-advice-on-health-spending-still-ignored-50-years-later/

Reid et al (2023) Implications of new proposed legislation on the credentialing and employment of healthcare workers in Ontario. https://www.osler.com/en/resources/regulations/2023/implications-of-new-proposed-legislation-on-the-credentialing-and-employment-of-healthcare-workers-i

Riley et al (2012) Health disparities: gaps in access, quality and affordability of medical care. Trans Am Clin Climatol Assoc 123:167–174

Rodrigues (2023) Public-private partnerships model applied to hospitals – a critical review. Healthcare 11(12):1723

Ross University, School of Medicine (2021) US vs. Canadian Healthcare: what is the difference? https://medical.rossu.edu/about/blog/us-vs-canadian-healthcare

Shah et al (2018) Achieving universal health coverage through public private partnerships: a study of trends in the public private partnerships in India. IJHSR 8(10):254–261

Shoppers Drug Mart (2023) Shoppers drug mart applauds expanded scope of practice for Ontario Pharmacists. https://corporate.shoppersdrugmart.ca/en/shoppers-drug-mart-applauds-expanded-scope-of-practice-for-ontario-pharmacists/

Statistics Canada (2021) Ethnocultural and religious diversity - 2021 Census promotional material. https://www.statcan.gc.ca/en/census/census-engagement/community-supporter/ethnocultural-and-religious-diversity

Statistics Canada (2022) Population growth in Canada's rural areas, 2016 to 2021. https://www12.statcan.gc.ca/census-recensement/2021/as-sa/98-200-x/2021002/98-200-x2021002-eng.cfm

Tasker (2024) Massive new survey finds widespread frustration with access to primary health care. https://www.cbc.ca/news/politics/primary-care-canada-10-000-canadians-report-1.7125990

The Canadian Press (2024) Canada ranks last in primary health care access among 10 wealthy countries: report. https://www.thestar.com/life/health-wellness/canada-ranks-last-in-primary-health-care-access-among-10-wealthy-countries-report/article_afd54669-5cd5-5dd1-b193-d95c98203897.html

U.S. Department of Health and Human Services, Digital Communications Division (2022) What's the difference between Medicare and Medicaid? https://www.hhs.gov/answers/medicare-and-medicaid/what-is-the-difference-between-medicare-medicaid/index.html

UK Trade & Investment (2013) Public Private Partnerships. https://assets.publishing.service.gov.uk/media/5a7c7835e5274a559005a152/07_PPP_28.11.13.pdf

Watts et al (2023) Ontario proposes new legislation for private health facilities and surgical centres. https://www.osler.com/en/resources/regulations/2023/ontario-proposes-new-legislation-for-private-health-facilities-and-surgical-centres#

Weeks C (2024) SickKids lowers surgery backlog through successful new partnership with nearby hospitals. https://www.theglobeandmail.com/canada/article-toronto-sick-kids-hospital-backlogs/

World Health Organization (2008) The right to health. In: The Human Rights Fact Sheet Series. Office of the United Nations High Commissioner for Human Rights, United Nations Office. https://www.ohchr.org/sites/default/files/Documents/Publications/Factsheet31.pdf

World Health Organization (2017) World Bank and WHO: half the world lacks access to essential health services, 100 million still pushed into extreme poverty because of health expenses. https://www.who.int/news/item/13-12-2017-world-bank-and-who-half-the-world-lacks-access-to-essential-health-services-100-million-still-pushed-into-extreme-poverty-because-of-health-expenses

World Health Organization (2023a) Billions left behind on the path to universal health coverage. https://www.who.int/news/item/18-09-2023-billions-left-behind-on-the-path-to-universal-health-coverage

World Health Organization (2023b) New WHO report lays out concrete actions for governments to optimize public–private partnerships for health. https://www.who.int/europe/news/item/26-01-2023-new-who-report-lays-out-concrete-actions-for-governments-to-optimize-public-private-partnerships-for-health

World Health Organization Regional Office for Europe (2023) Public–private partnerships for health care infrastructure and services: policy considerations for middle-income countries in Europe. In: Health financing policy papers. WHO Barcelona Office for Health Systems Financing. https://iris.who.int/bitstream/handle/10665/365603/9789289058605-eng.pdf?sequence=1&isAllowed=y

Your Health Act, 2023, SO 2023, c 4

Jennifer D. Suess oversees all aspects of RioCan's corporate legal strategy and related execution, governance, ESG, insurance, Board and related Committee matters and acts as Corporate Secretary. As a key member of RioCan's senior management team and advisor to the Board, Jennifer is actively engaged in formulating and managing the strategy for the organization. Jennifer is RioCan's first-ever General Counsel, having joined the company in 2017 after serving as a corporate/securities lawyer with a sub-specialty in the healthcare sector at leading multi-national companies including Johnson & Johnson and Novartis. Jennifer puts her deep experience and passion to work in order to provide counsel and strategic guidance that helps propel RioCan toward its vision. In addition to her work at RioCan, since 2022, Jennifer is also a member of the Board of Directors of The Hospital for SickKids Foundation where she sits on the Nominating & Governance and Compensation & Resource Management Committees, and previously was a member of the Development Committee. In 2022, Jennifer was Valedictorian and completed her ICD.d certificate in the Director Education Program at the Rotman School of Management, University of Toronto. In 2023, she received her Competent Boards ESG Certification from Competent Boards. Prior to joining RioCan, Jennifer began her legal career at Blake, Cassels & Graydon LLP and Goodmans LLP. She was also the inaugural Head of Legal & Compliance at Alcon Canada Inc., a Novartis company, where she acted as Corporate Secretary & Director, Head of Human Resources, Head of Communications and Chief Privacy Officer. Called to the Bar in 2003, Jennifer achieved her Law degree at the University of Windsor and holds a Bachelor of Science degree from the University of Toronto. She has received many awards including being named as one of Canada's Top 25 Most Influential Lawyers in 2024, receiving the Lexpert 2021 Female Trailblazer Excellence Award, the 2020 Chambers Diversity & Inclusion North America Award and the 2019 Canadian General Counsel ESG Award for Excellence. On January 1st, 2025, Jennifer was appointed to the Order of Ontario, the province's highest civilian honour.

Indigenous Law and Sustainability

Sandy Carpenter

Abstract

This chapter follows one aspect of Indigenous law, the "duty to consult", which many people consider the most important development in Indigenous law in the past 30 years, over part of the career of a lawyer who has practiced in, and through, that development. It then offers a perspective on the contribution of the duty to consult to the sustainability of Indigenous Nations and people in Canada.

The views expressed are solely those of the lawyer and remain open to change.

1 Introduction

I didn't know either of the words Indigenous or sustainability for a long time. However, given my chosen career as a lawyer, and then the focus of my practice, I probably learned these before most people.

2 Prince George and Indians

I grew up in Prince George, near the middle of British Columbia.

Prince George is at the confluence of the Fraser and Nechako Rivers. More recently, it's also the confluence of Highways 16 and 97, the primary north/south, east/west corridors in central British Columbia.

West of Prince George, Highway 16 is now more infamously known as the Highway of Tears, where dozens of Indigenous women and girls have gone missing

S. Carpenter (✉)
P2 Regulatory Solutions, Calgary, AB, Canada
e-mail: sandycarpenter@regulatoryandindigenouslaw.com

A. McLaughlin, C. Chell (eds.), *Building the Case: How Lawyers are Leading on Sustainability*, https://doi.org/10.1007/978-3-031-88455-9_10

or been found murdered. When I lived in Prince George, nobody I knew was talking about Missing and Murdered Indigenous Women and Girls.[1]

I met my first Indians in Prince George. At the time, people used different names for who I hope most people now refer to as Indigenous people. Most of these were, at best, derogatory. I always felt that I should just refer to them as Indians. It seemed to be the least derogatory of the choices I was aware of.

At the time, I didn't know that Indigenous people were called Indians in Canada's Constitution.[2] As far as I knew, nobody referred to Indians as Aboriginal, Indigenous, First Nation, Inuit, Métis, or any of the hundreds of names that Indigenous people call themselves.

In hindsight, meeting an Indian in Prince George wasn't unusual. What is now Prince George is on the unceded territory of the Lheidli T'enneh First Nation, "The People from the Confluence of the River",[3] and Indigenous people have occupied the area for thousands of years.[4]

3 Other Things I Didn't Know

The Canadian government passed the first *Indian Act* in 1876.[5] It consolidated and expanded on earlier colonial laws about "Indians", including legislating who was an Indian.[6] As then Prime Minister John A. Macdonald described the *Indian Act*:

> The great aim of our legislation has been to do away with the tribal system and assimilate the Indian people in all respects with the other inhabitants of the Dominion as speedily as they are fit to change.[7]

Over the next decades, amendments to the *Indian Act* reinforced this policy, such as:

[1] A National Inquiry into Missing and Murdered Indigenous Women and Girls took place in Canada from 2016 to 2019 culminating in a two-volume report. National Inquiry into Missing and Murdered Indigenous Women and Girls (2019): Reclaiming Power and Place: The Final Report of the National Inquiry into Missing and Murdered Indigenous Women and Girls.

[2] *An Act of the Imperial Parliament for the Union of Canada, Nova Scotia and New Brunswick* (aka the *British North America Act, 1867*), 30 & 31 Vict. c. 3. Section 91(24) of what is now known as the *Constitution Act, 1867* provides that the federal government has legislative jurisdiction over "Indians, and Lands reserved for the Indians".

[3] Lheidli T'enneh (2024) Our Story. https://www.lheidli.ca/about/our-story/.

[4] City of Prince George (2024) History of Prince George. https://www.princegeorge.ca/community-culture/heritage/history-prince-george.

[5] *Indian Act*, SC 1876, c. 18.

[6] The *Indian Act* was not only racially directed but misogynistic. Any woman married to a "male person of Indian blood belonging to a particular band" was an Indian but an "Indian woman marrying anyone other than an Indian or non-treaty Indian" ceased to be an Indian (section 3). Only males over the age of 21 could vote in band elections (section 61).

[7] Canada, Sessional Papers, "Return to an Order of the House of Commons, May 2, 1887, (no. 20b) at 37.

- In 1880, granting the power to prohibit traditional Indigenous governance structures and to impose an electoral system of governance;[8]
- In 1884, outlawing traditional ceremonies;[9]
- In 1894, granting the power to establish industrial schools or boarding schools for Indians, and the committal of Indian children under the age of 16 years to such schools;[10]
- In 1911, allowing for the expropriation of reserve lands;[11] and
- In 1940, prohibiting the sale of furs and wild animals except in accordance with regulations.[12]

These are only a few examples.

4 The *Constitution Act, 1982* and Section 35

The *Constitution Act, 1982* was passed the year I started law school. It "patriated" Canada's Constitution, constitutionalized a Charter of Rights and Freedoms, and tried to clarify the division of powers between the federal and provincial governments over natural resources.[13]

It also included section 35, which reads in part:

> **35.** (1) The existing aboriginal and treaty rights of the aboriginal peoples of Canada are hereby recognized and affirmed.
>
> (2) In this Act, aboriginal peoples of Canada includes the Indian, Inuit and Métis peoples of Canada. [...][14]

For some, section 35 represented the culmination of a long struggle for the constitutional recognition of Aboriginal rights.[15] For others, it didn't seem as significant. Despite taking Constitutional Law, Natural Resources Law, Forestry Law, Mining

[8] S.C. 1880, c. 28, s. 72.

[9] SC 1884, c. 28, s. 3.

[10] SC 1894, c. 32, s. 11. Further amendments provided for the power to pass regulations allowing for the "arrest and conveyance to school, and detention there, of truant children and of children who are prevented by their parents or guardians from attending: and such regulations may provide for the punishment, upon summary conviction, by fine or imprisonment, or both, of parents and guardians, or persons having the charge of children, who fail, refuse or neglect to cause such children to attend school."

[11] SC 1911, c. 14, s. 1, 2.

[12] SC 1940-41, c.19, s. 1.

[13] Schedule B to the *Canada Act 1982* (UK), 1982, c.11.

[14] Notwithstanding the express recognition of Métis rights in section 35, it wasn't until 2016 that the Supreme Court of Canada confirmed that Métis people fall within the definition of "Indians" under section 91(24) of the *Constitution Act, 1867* (*Daniels v. Canada (Indian Affairs and Northern Development*, 2016 SCC 12). Inuit people were recognized as falling under section 91(24) in 1939 (*Reference Re Eskimos*, [1939] SCR 104).

[15] *R. v. Sparrow*, [1990] SCR 1075, p. 1105 (*Sparrow*).

Law, and Administrative Law (among others) in law school, I don't remember anybody mentioning it.

I was taught that Her Majesty the Queen in Right of British Columbia owned most of British Columbia, including all the forest, mineral, oil and gas, and water resources (except those that had been transferred to private interests before the government realized how valuable they were). I was also taught, notwithstanding what I learned about trespass in Real Property Law that, through the provincial government's legislative powers, it was legal for a "free miner" to enter someone's property and acquire an interest in the minerals under it—without ever giving notice to the "owner" of the property—and without any thought about how this could affect Indigenous interests.[16]

5 *Sparrow* and Soft Tissue Injuries

On May 25, 1984, Ronald Sparrow, a member of the Musqueam Nation, was caught fishing with a net longer than permitted under the Musqueam's fishing licence. Mr. Sparrow was charged with violating the *Fisheries Act*.[17]

Mr. Sparrow took a hard line in his defence. He argued that he was exercising an aboriginal right to fish and, since his rights were constitutionally recognized and affirmed under section 35, the government had no power to regulate them.[18]

The federal government took an equally hard line. Since Mr. Sparrow's right to fish had been regulated prior to section 35 being included in the *Constitution*, any traditional right to fish had been extinguished and Mr. Sparrow only had the rights that were given to him by the government.[19]

The Supreme Court of Canada rejected both positions and forged a middle ground. The legislation regulating Mr. Sparrow's rights had not shown a "clear and plain" intention to take them away; therefore, Mr. Sparrow's rights were alive and well.[20] However, that did not mean they were immune from regulation. The government could still regulate Mr. Sparrow's rights if it could "justify" doing so—one aspect of which was whether the government had "consulted" the Indigenous group about the proposed "infringement".[21]

"Consultation" was only one non-exclusive factor in the *Sparrow* justification test; however, the groundwork was laid. If the government wanted to do something that could have a negative impact on Indigenous rights, it had to actually talk to the Indigenous group(s) about what the government was thinking.

I had been practicing for almost four years when *Sparrow* came out. In law school, I thought I wanted to be a real estate lawyer. That was quickly displaced by

[16] *Mineral Act*, RSBC 1979, c. 259, s. 6.

[17] *Sparrow*, p. 1083.

[18] *Sparrow*, pp. 1086-1087 and 1102.

[19] *Sparrow*, pp. 1087 and 1095–1097.

[20] *Sparrow*, p. 1099.

[21] *Sparrow*, pp. 1109–1110; 1114–1119.

the polycentric real life fix of regulatory law rather than the post-facto application of "black letter law".

Unfortunately, there wasn't a lot of regulatory work at the time and, while I was aware of *Sparrow*, it wasn't particularly relevant to a young lawyer trying to keep himself busy defending personal injury claims and collecting debts.

6 Floating Coffins

In *Rio Tinto*, the Supreme Court of Canada set out a brief history of one of the largest resource development projects that has taken place in Canada:

> [3] In the 1950s, Alcan (now Rio Tinto Alcan) dammed the Nechako River in northwestern British Columbia for the purposes of power development in connection with aluminum production. The project was one of huge magnitude. It diverted water from the Nechako River into the Nechako Reservoir, where a powerhouse was installed for the production of electricity. After passing through the turbines of the powerhouse, the water flowed to the Kemano River and on to the Pacific Ocean in the west. The dam affected the amount and timing of water flows into the Nechako River to the east, impacting fisheries on lands now claimed by the CSTC First Nations. Alcan effected these water diversions under Final Water Licence No. 102324 which gives Alcan use of the water on a permanent basis.[22]

Alcan's original aluminum production facilities did not require all the water rights it had acquired. Therefore, Alcan still needed to be able to release the water that it didn't need to the east. This was to be done using the "Cheslatta Spillway", releasing about half the historic volume of the upper Nechako River into and through the previously separate Cheslatta watershed where, after passing over Cheslatta Falls, it would rejoin the original location of the Nechako River.[23]

Inconveniently, the Cheslatta watershed was the home of the Cheslatta Carrier Nation. Early in 1957, shortly before Alcan would have to begin releasing water from the spillway, the federal Department of Indian Affairs told the Cheslatta that they needed to leave their homes. Indian Affairs then burned their homes so they couldn't return.

A short time later, Alcan started releasing water. The now, much larger, flows eroded a bluff where the Cheslatta's graveyard was located, washing grave boxes and the Cheslatta's ancestors downstream.[24]

---

[22] *Rio Tinto Alcan Inc. v. Carrier Sekani Tribal Council*, 2010 SCC 43, para. 3 (*Rio Tinto*). The Supreme Court was actually wrong on the last point. Alcan acquired the right to dam and divert the upper Nechako River under a 1950 Agreement with the Province of British Columbia and Conditional Water Licence No. 19847 dated December 29, 1950.

[23] Unlike many dams, the Kenney Dam, which blocked the flow of the upper Nechako River, is a solid rock-fill dam and does not have any way of releasing water below it. The Cheslatta watershed was previously a minor tributary of the Nechako disconnected from the large chain of lakes forming the headwaters of the Nechako River. The Cheslatta watershed's natural average annual flows were in the order of 1 cubic meter/second.

[24] After Alcan started releasing water into the Cheslatta system, the average annual flows through the Cheslatta system have been around 94 cubic meters/second.

I first heard the Cheslatta's story in 1995. The BC government had asked the BC Utilities Commission to conduct an inquiry into Alcan's plan to make use of (some of) its remaining rights. I was co-counsel for Alcan, having hopefully left car crashes and unpaid debts behind.

During an early session of the inquiry, the proceedings were disrupted by banging on the building where the session was taking place. When we went outside, the building was surrounded by Indigenous people. Some were elderly; many were young.

My reaction was mixed. I've never thought I can put myself in an Indigenous person's shoes, but I thought I could understand how people who had been so harshly treated would still be angry. But why would you want to pass on that anger onto young children?

Not long after, I changed my mind. If I had been removed from my home, forced to travel with my only possessions on horseback or wagon through rutted mud tracks to a place I wasn't from, had my home and remaining possessions burned, told I could now farm to sustain myself, and had my ancestors' graves wash away, I think I'd want my children and grandchildren to know where they came from as well.

The inquiry, which ultimately took 87 hearing days, and its aftermath, changed my life.[25] Professionally, I had levelled up. Personally, I lost a relationship and gained 40 pounds. Through the Cheslatta and others, I started to understand the legacy and continuing impacts of "development" on Indigenous people.[26]

7 A Right to the Land Itself

In 1997, the Supreme Court of Canada dropped *Delgamuukw*.[27]

Thirteen years earlier, Delgamuukw and 50 or so other Hereditary Chiefs of the Gitksan and Wet'suwet'en Nations had sued the Province of British Columbia for a declaration of ownership and jurisdiction over 58,000 square km of land.[28] Predictably, British Columbia counterclaimed for a declaration that the Chiefs had no right to or interest in the claimed territory. Alternatively, the Province argued the Chiefs should have sued the Government of Canada for compensation.

---

[25] Including being unceremoniously thrown into the world of sustainability, and getting to write an early Corporate Sustainability Report.

[26] Dozens of Indigenous communities have been relocated in Canada. Some of these, including the relocation of the Cheslatta, were discussed by the Royal Commission on Aboriginal Peoples. Royal Commission on Aboriginal Peoples (1996): Report of the Royal Commission on Aboriginal Peoples: Volume 1, Looking Forward, Looking Back, Chapter 12.

[27] *Delgamuukw v. British Columbia*, [1997] 3 S.C.R. 1010 (*Delgamuukw*).

[28] "The fundamental premise of both the Gitksan and the Wet'suwet'en peoples is that they are divided into clans and Houses. Every person born of a Gitxsan or Wet'suwet'en woman is automatically a member of his or her mother's House and clan. There are four Gitksan and four Wet'suwet'en clans, which are subdivided into Houses. Each House has one or more Hereditary Chief as its titular head [...]" (*Delgamuukw*, para. 12).

After 374 days of evidence, the trial judge dismissed the Chiefs' claims.[29]

On appeal, the Chiefs switched from claiming ownership and jurisdiction in each of their "Houses" to claiming Aboriginal title and self-government on behalf of their respective Nations. Notwithstanding this refinement, their appeal was dismissed.[30]

The Chiefs appealed again to the Supreme Court of Canada. Like in *Sparrow*, the Court refused to accept either the Chiefs' or the Province's positions. Aboriginal title does not mean that Aboriginal people can use the land however they choose.[31] However, nor is it simply a "bundle" of rights to engage in otherwise recognized rights, or the use and occupation of lands to carry out those rights.[32] Aboriginal title is *a right to the land itself.*

While the Supreme Court's definition of Aboriginal title allows the land to be used for activities that don't need to be Aboriginal rights, the land can't be used for purposes that are inconsistent with the Nation's attachment to the land.[33] The lands also cannot be transferred, sold or surrendered to anyone other than the Crown.[34]

From an Indigenous/Aboriginal law perspective, *Delgamuukw* turned the existing legal paradigm on its head. It was finally clear that the "black letter law" that the Crown owned everything (except for what it had previously transferred), simply wasn't true.[35]

Sadly, at least to the Gitksan and Wet'suwet'en, while the Court finally recognized the theoretical existence of Aboriginal title, the Chief's claims were sent back to trial.

To early duty to consult practitioners, *Delgamuukw* was also significant for synthesizing the overarching law of consultation, as it had developed since *Sparrow*:

> [168] [...] There is always a duty of consultation. Whether the aboriginal group has been consulted is relevant to determining whether the infringement of aboriginal title is justified, in the same way that the Crown's failure to consult an aboriginal group with respect to the terms by which reserve land is leased may breach its fiduciary duty at common law: *Guerin*. The nature and scope of the duty of consultation will vary with the circumstances. In occasional cases, when the breach is less serious or relatively minor, it will be no more than a duty to discuss important decisions that will be taken with respect to lands held pursuant to aboriginal title. Of course, even in these rare cases when the minimum acceptable standard is consultation, this consultation must be in good faith, and with the intention of substantially addressing the concerns of the aboriginal peoples whose lands are at issue. In most cases, it will be significantly deeper than mere consultation. Some cases may even require the full consent of an aboriginal nation, particularly when provinces enact hunting and fishing regulations in relation to aboriginal lands.

[29] [1991] 3 WWR. 97 (BCSC).

[30] [1993] 5 WWR 97 (BCCA).

[31] Aside from government regulation and expropriation.

[32] *Delgamuukw*, para. 110.

[33] *Delgamuukw*, paras. 111 and 117 to 132.

[34] *Delgamuukw*, para. 113.

[35] In *Calder et al. v. Attorney-General of British Columbia*, 1973 CanLII 4 (SCC), three judges of the Supreme Court found in favour of Aboriginal title, three found against, and one dismissed the case on a procedural basis.

By the time *Delgamuukw* was released, I had moved to Alberta, where resource development (mostly oil and gas) was almost exclusively in areas of proven Treaty rights. While *Delgamuukw* addressed the required consultation before governments could infringe Aboriginal title, it was clear the same standard would apply to Treaty rights. *Badger* had already confirmed that *Sparrow's* justification test also applied to treaty rights.[36]

8 Everything Everywhere All at Once

Today, when people think of the duty to consult, they think of *Haida* (discussed below). However, as is clear from the above, (some) governments were consulting Indigenous people long before *Haida* was decided in 2004.

Originally consultation took place, or didn't, in the context of regulatory restrictions that impacted Indigenous rights, such as Mr. Sparrow's fishing rights.

However, First Nations quickly recognized that the same principles could be applied to resource development permits or licences.

In *Ryan*, in 1994, the Gitskan challenged a District Forest Manager's decision approving logging in an area the Gitxsan claimed as their traditional territory.[37] In *Metecheah*, in 1997, the Halfway River First Nation challenged the approval of a cutting permit.[38]

In 1998, I started a (short-lived) tenure as outside counsel for the new BC Oil and Gas Commission (now the BC Energy Regulator). By then, the "OGC" already had dedicated liaisons and consultation agreements in place, including capacity funding, with each of the Treaty 8 Nations in northeast BC.

9 You Always Remember Your First

The first duty to consult court case I argued was *Apsassin*. I had been involved in numerous consultations, negotiations, and agreements with Indigenous groups, but none had ended up in court.

I acted for Vintage Petroleum. Vintage had the right (from Her Majesty) to explore for oil and gas in a largely undeveloped area in Treaty 8. It wanted to drill an exploratory well to see if it was worth drilling other wells.

Both the OGC and Vintage consulted with the potentially impacted First Nations, and Vintage committed to a number of measures to minimize the impact of its proposed activities.

The Saulteau First Nation took the position there needed to be a cumulative effects assessment of the potential impacts of future development prior to any drilling taking place. Ultimately, the OGC approved Vintage's proposed well without a

[36] *R. v. Badger*, [1996] 1 SCR 771, paras. 37, 73, 85, and 96 to 97 (*Badger*).

[37] *Ryan v. Shultz* (January 25, 1994), Smithers Registry, No. 7855 (S.C.) (*Ryan*)

[38] *Metecheah v. Ministry of Forests*, 1997 CanLII 2719 (BCSC) (*Halfway River*).

cumulative effects assessment. The Saulteau challenged the permit. I was retained after consultation had taken place.

In 2003, you could still read all of the duty to consult cases that had been decided, so I did. I thought that it would be good to clearly set out the (then) law on "consultation" since there was no guarantee the judge would be familiar with the area (and still isn't).

I also felt that I needed to find a tiebreaker. It was clear that consultation had taken place but had ended up with different positions. Who was right?

Like many early consultation cases, *Nikal* arose out of an alleged regulatory offence. Fisheries regulations required everyone to get a fishing licence. Mr. Nikal, a member of the Wet'suwet'en, didn't have one and he was charged with fishing without a licence. Mr. Nikal argued that making him get a licence infringed his rights.

Mr. Nikal won.[39] A majority of the Supreme Court found that the conditions on his licence, had he actually got one, would have been an unjustified infringement of his right to fish. However, as usual, Mr. Nikal didn't win on his more sweeping argument that the mere requirement to get a licence infringed his rights. As Mr. Justice Cohen described *Nikal* in *Apsassin*:

> [102] In ***R. v. Nikal***, 1996 CanLII 245 (SCC), [1996] 1 S.C.R. 1013, at paras. 91 to 93, the Court held that while aboriginal and treaty rights are given significant weight in this framework, as indicated by the Court, these rights are not absolute and there needs to be some balance between other legitimate legislative objectives and the rights of aboriginal peoples:
>
> With respect to licensing, the appellant takes the position that once his rights have been established, anything which affects or interferes with the exercise of those rights, no matter how insignificant, constitutes a *prima facie* infringement. It is said that a licence by its very existence is an infringement of the aboriginal right since it infers that government permission is needed to exercise the right and that the appellant is not free to follow his own or his band's discretion in exercising that right.
>
> This position cannot be correct. It has frequently been said that rights do not exist in a vacuum, and that the rights of one individual or group are necessarily* limited by the rights of another. The ability to exercise personal or group rights is necessarily limited by the rights of others. The government must ultimately be able to determine and direct the way in which these rights should interact. Absolute freedom in the exercise of even a *Charter* or constitutionally guaranteed aboriginal right has never been accepted, nor was it intended. Section 1 of the *Canadian Charter of Rights and Freedoms* is perhaps the prime example of this principle. Absolute freedom without any restriction necessarily infers a freedom to live without any laws. Such a concept is not acceptable in our society. [...] [40]

Nikal should be required reading for anyone engaged in consultation – and also some of the non-Indigenous people who seem to constantly think their rights are being infringed by anything and everything.

---

[39] *R. v. Nikal*, [1996], 1 SCR 1013.

[40] .*Apsassin et al v. BC Oil and Gas et al*, 2004 BCSC 92, para. 102; see also para. 146 (*Apsassin*).

While I won *Apsassin*, I also learned an important lesson. The well never got drilled.

10 *Haida*

On November 18, 2004, the Supreme Court of Canada released *Haida*.[41] As alluded to, *Haida* is widely regarded as the leading case on governments' obligation to consult with Indigenous people in Canada.

The Court expressed the issue this way:

> This brings us to the issue before this Court. The government holds legal title to the land. Exercising that legal title, it has granted Weyerhaeuser the right to harvest the forests in Block 6 of the land. But the Haida people also claim title to the land — title which they are in the process of trying to prove — and object to the harvesting of the forests on Block 6 as proposed in T.F.L. 39. In this situation, what duty if any does the government owe the Haida people? More concretely, is the government required to <u>consult</u> with them about decisions to harvest the forests and to <u>accommodate</u> their concerns about what if any forest in Block 6 should be harvested *before they have proven their title to land and their Aboriginal rights*?
>
> The stakes are huge. The Haida argue that absent consultation and accommodation, they will win their title but find themselves deprived of forests that are vital to their economy and their culture. Forests take generations to mature, they point out, and old-growth forests can never be replaced. The Haida's claim to title to Haida Gwaii is strong, as found by the chambers judge. But it is also complex and will take many years to prove. In the meantime, the Haida argue, their heritage will be irretrievably despoiled.
>
> The government, in turn, argues that it has the right and responsibility to manage the forest resource for the good of all British Columbians, and that until the Haida people formally prove their claim, they have no legal right to be consulted or have their needs and interests accommodated.[42] (emphasis added)

The Court found, dismissing the Province's argument that consultation was only required for proven rights, that consultation was also required on "asserted" rights. To accommodate the lack of certainty on whether actual rights were in issue, the Crown's required response was a product of both the strength of an Aboriginal group's claim and the degree of potential impact. A higher strength of claim and potential impact equals a higher response. A lower strength of claim and potential impact equals a lower response. There can be other mixtures in the middle.[43]

Haida also confirmed that industry did not have an independent duty to consult First Nations—but then found that industry could be delegated certain aspects of consultation.[44]

---

[41] *Haida Nation v. British Columbia (Minister of Forests)*, 2004 SCC 73 (*Haida*).

[42] *Haida*, paras. 6 to 8.

[43] *Haida*, paras. 31 to 50.

[44] *Haida*, paras. 52 to 56.

Sparrow had found that Indigenous rights were alive and well, but also recognized, in our necessary act of balancing rights, that Indigenous rights are not absolute and can be infringed so long as the government can justify the infringement, including consultation.

Delgamuukw told us that Aboriginal title was there, it was just waiting for someone to prove it. While not strictly necessary given the Court had sent the Gitksan and Wet'suwet'en's claims back to trial, the Court still took the opportunity to set out the broad principles of consultation.

Haida completed the circle. Regardless of whether an Indigenous right was proven or not, the government needed to consult with, and potentially accommodate, the Indigenous rights holder before making a decision that could impact that right. Importantly, while not imposing an obligation directly on industry, it also created a vehicle where industry could be directly involved in these consultations/negotiations.

11 Not So Cross Cross-Examinations

In 2010, I found myself in an awkward situation.

I had acted as external counsel to get a large project approved.[45] The decision-maker's decision was then challenged by a number of First Nations and, along with a similar finding in a different proceeding, was overturned by the BC Court of Appeal.[46]

The proponent in the other case decided to try to appeal to the Supreme Court of Canada. This resulted in *Rio Tinto* (introduced above), which upheld the Court of Appeal's decision that the decision maker in that case, and other regulatory tribunals with the necessary powers, are required to assess the adequacy of consultation whenever they consider a course of action that could impact Indigenous rights.[47]

However, sometimes projects just need to get built, and can't wait until all of the possible legal proceedings are exhausted. This resulted in, to the best of my knowledge, the first regulatory proceeding in Canada whose sole function was to determine whether Indigenous consultation had been adequate.

Cross-examinations are an effective tool for ascertaining the "truth". As a result, they are a fundamental part of the Canadian justice system - and have been adopted by some regulatory bodies as part of their toolkit. Cross-examinations are intense and can be confrontational. In this case, the decision-maker decided that cross-examinations would be part of their process.

On the spectrum, I'd describe myself as a quiet cross-examiner. I generally try to get a witness to agree with my more than reasonable propositions rather than being immediately aggressive, as some lawyers feel they need to do. However, in every cross-examination I had done, I had always been ready to engage in some form of

[45] British Columbia Utilities Commission Order C-4-08 dated August 5, 2008.

[46] *Kwikwetlem First Nations v. British Columbia (Utilities Commission)*, 2009 BCCA 68.

[47] Which the Court ultimately did. See *Rio Tinto*, paras. 55 to 63.

direct confrontation to get the witness to agree with my question, or be unbelievable in disagreeing.

I knew it was a bad idea to engage in direct confrontation with the Indigenous witnesses before the decision-maker. My client had been reasonable and respectful in their approach and if I engaged in the type of confrontation that can sometimes occur during cross-examination, I didn't think it would be consistent with how my client had behaved - and may cause the decision-maker to second guess my client's story.

Sixteen witnesses ultimately appeared for the Nations and representative groups. Seven were Chiefs. I don't remember all of my cross-examinations. I remember one started with asking if I was pronouncing Nlaka'pamux correctly. I repeated Nlaka'pamux until the witness confirmed I had it right.

Thirteen days of cross-examination and a 258-page decision later, the decision-maker (with a dissent) found that my client had consulted adequately with a number of First Nations and inadequately, on specific topics, with others.[48]

My client responded to the areas where consultation was found to be inadequate and the Commission accepted that consultation had been adequate for all of the First Nations.

At least I don't need to do all of my cross-examinations that way.

12 Twenty Years Later

On November 18, 2024, it was twenty years since *Haida* was decided and longer since the earliest consultations began.

Each consultation has its own issues and stories. However, the prescience of *Sparrow* and *Delgamuukw*, coupled with the principles and framework in *Haida* have been remarkably resilient.

The trend has been to colour in the spaces, not change the lines, but that doesn't mean there haven't been significant developments. In *Gitxaala*, the BC Supreme Court recently found there needs to be consultation with potentially affected First Nations before a free miner can stake a mineral claim, overturning over a hundred years of practice, and the starting point for everything I learned in my Mining Law class.[49]

There has also been a trend to institutionalize consultation. In 1994, the first BC *Environmental Assessment Act*,[50] which was remarkably ahead of its time, included ten references to "First Nations", many of these in the same sections. The current BC *Environmental Assessment Act* never refers to First Nations, but it has 87 references to Indigenous Nations, Indigenous peoples, Indigenous knowledge,

[48] In the Matter of British Columbia Transmission Corporation Court of Appeal Reconsideration of the Interior to Lower Mainland Transmission Project Decision, February 3, 2011.

[49] *Gitxaala v British Columbia (Chief Gold Commissioner)*, 2023 BCSC 1680 (*Gitxaala*).

[50] SBC 1994, c. 35.

Indigenous rights,[51] and, recently, an *Environmental Assessment Dispute Resolution Facilitators Regulation*, providing access to a dispute resolution process for Indigenous Nations.[52]

However, at least from my perspective, the greatest trend is the movement of discussions away from narrowly focused consultation on avoiding or mitigating potential impacts to proven or asserted Indigenous rights and title to broader negotiations about partnerships and economic opportunities. While this latter trend clearly has its roots in the duty to consult, and particularly in attempting to mitigate the risk that consultation might be found to be inadequate, the trend to negotiated outcomes and broader relationship measures has also been amplified by other events such as the Truth and Reconciliation Commission's Calls to Action,[53] the ESG movement, and the adoption of the United Nations Declaration on the Rights of Indigenous Peoples.[54]

This is not to say that "pure" consultation has gone away. To the contrary, the constitutional requirement is still firmly in place and remains the trigger for starting, maintaining, finalizing, and continuing for as long as necessary discussions and resolutions.

13 So, What About Sustainability?

As I've said, I can't put myself in the shoes of an Indigenous person. I'm also not a sustainability expert. I've practiced what I now refer to as Indigenous law since just after *Sparrow* and regulatory law for longer than that. My reflections are based on that experience.

If the question is whether the introduction of section 35 and the development of the duty to consult has contributed to the sustainability of Indigenous rights and peoples in Canada in comparison to what occurred prior to section 35, my answer is an unequivocal yes. Of course, that's a really low bar.

If the question is whether the duty to consult has contributed to the sustainability of Indigenous rights and people since its conception, to me, it depends. For some communities, the duty to consult has provided the opportunity for a fundamentally different future. For others, section 35's protections and the duty to consult clearly haven't been enough. As stated in *Yahey*:

> I conclude that the extent of the lands taken up by the Province for industrial development (including the associated disturbances, impacts on wildlife, and impacts on Blueberry's

[51] SBC 2018, c. 51.

[52] OIC No. 465.

[53] Truth and Reconciliation Commission of Canada (2015): Honouring the Truth, Reconciling for the Future Summary of the final report of the Truth and Reconciliation Commission of Canada. Please note that the Truth and Reconciliation Commission published a number of detailed reports. These reports are all available through the link above.

[54] United Nations, Office of the High Commission for Human Rights (2007): United Nations Declaration on the Rights of Indigenous Peoples.

> way of life), means there are *no longer sufficient and appropriate lands in Blueberry's territory to allow for the meaningful exercise by Blueberry of its treaty rights*. The cumulative effects of industrial development authorized by the Province have significantly diminished the ability of Blueberry members to exercise their rights to hunt, fish and trap in their territory as part of their way of life and therefore constitute an infringement of their treaty rights. The Province has not justified this infringement.[55] (emphasis added)

Many, if not most, of the impacts to the Blueberry First Nation have taken place since the duty to consult has been in place.

If the question is whether more needs to be done to sustain Indigenous rights and peoples in Canada, my answer is an unequivocal yes. While I've been writing this chapter a number of headlines have crossed my path. While there have been positive stories, this is a small but representative sample of the others:

- A "Gut-Wrenching" Snapshot of First Nations Health (August 27, 2024);[56]
- Canadian killer given life sentences for four murders (August 28, 2024);[57] and
- NDP MP says action needed now to "save Indigenous people's lives' after 6 killings by police (September 18, 2024).[58]

I didn't know the words Indigenous or sustainability for a long time. While I hear both almost daily now, I still don't hear them used together very often.

Sandy Carpenter is a Co Founder of P2 Regulatory Solutions and is widely recognized as one of Canada's leading regulatory and Indigenous lawyers. Sandy has practiced administrative, environmental and Indigenous law across a wide range of industries, jurisdictions and time, and has written and spoke about a broad range of legal issues.

[55] *Yahey v. British Columbia*, 2021 BCSC 1287 (*Yahey*).

[56] The Tyee (2024) https://thetyee.ca/News/2024/08/27/Gut-Wrenching-Snapshot-First-Nations-Health/.

[57] British Broadcasting Corporation (2024) https://www.bbc.com/news/articles/cj08n3zd7zqo.

[58] Canadian Broadcasting Corporation (2024) https://www.cbc.ca/news/indigenous/first-nations-police-deaths-government-1.7327312.

Social Sustainability and Workplace Law

Jillian Frank

Abstract

This chapter provides a brief history of Canadian workplace law and social sustainability and focuses on how employment lawyers can shape corporate social responsibility at the workplace. The author's own professional experiences and career as an employment lawyer are included as examples of building a career in social sustainability.

Social sustainability enables growth through the promotion and availability of new markets, business partners, and sources of innovation. Employment opportunities, workplace safety, and equity are core parts of social sustainability. Canadian laws recognize the significance of work in our communities and the importance of removing barriers to an individual's ability to earn a living and contribute to society through their work. Employment lawyers must know and consider these laws when advising clients—in other words, employment lawyers *are* social sustainability lawyers. And, as social expectations and working conditions continue to change, lawyers have an opportunity to influence the future of social sustainability initiatives. Employment lawyers play an important role in defining the line between voluntary social responsibility and mandatory legal obligations and will influence how our laws need to adapt in the face of artificial intelligence, economic transition, flexible work arrangements, and contract-based work.

J. Frank (✉)
KPMG Law LLP, Vancouver, BC, Canada
e-mail: jillianfrank@kpmg.ca

A. McLaughlin, C. Chell (eds.), *Building the Case: How Lawyers are Leading on Sustainability*, https://doi.org/10.1007/978-3-031-88455-9_11

1 Introduction: Social Sustainability and the Workplace

> When the Stranger says: 'what is the meaning of this city? Do you huddle close together because you love each other? What will you answer? 'We all dwell together to make money from each other'? or 'this is a community'.

– TS Eliot, *Choruses from the Rock*

> Development has to be more concerned with enhancing the lives we lead and the freedoms we enjoy.

– Amartya Sen, *Development as Freedom*

When lawyers talk about sustainability and ESG, they often refer to climate change, environmental protection laws and standards. But building a sustainable future goes beyond protecting our physical surroundings. It also requires consideration of basic rights and freedoms to health and wellbeing, equity, and stable work, which give people opportunities to thrive, contribute, and build communities. Environmental and economic problems are linked to social and political factors, including the status of women, the distribution of power, the protection of vulnerable groups, and participation in decision making.[1]

As an employment lawyer, I have a role to play in shaping the way Canadian laws enable workers to thrive in their communities. This chapter focuses on the link between social sustainability and practicing employment law in Canada. Working as a legal advocate in social sustainability doesn't necessarily mean becoming an international human rights advocate or working for the United Nations—Canadian workplace laws also play an important part in advancing social sustainability.

Social sustainability is based on the concept that poverty and inequality affect us all, and that our long-term ability to grow and thrive involves taking steps to enable communities to exist in a healthy, fair, and equitable way. Social sustainability enables growth through the promotion and availability of new markets, business partners, and sources of innovation.

Employment opportunities, workplace safety, and equity are core parts of social sustainability. The United Nations Sustainable Development Goals (UNSDGs) promote inclusive and sustainable economic growth and stable, productive jobs.[2] The United Nations Declaration of Human Rights (UDHR) recognizes rights and responsibilities relating to work, including that:

---

[1] See the Report of the World Commission on Environment and Development: Our Common Future (the Brundtland Report), March 20, 1987 at para 43, Our Common Future: Report of the World Commission on Environment and Development (un.org).

[2] United Nations Sustainable Development Goals, Goal 8. Economic Growth - United Nations Sustainable Development.

1. Everyone has the right to work, to free choice of employment, to just and favourable conditions of work and to protection against unemployment.
2. Everyone, without any discrimination, has the right to equal pay for equal work.
3. Everyone who works has the right to just and favourable remuneration ensuring for himself and his family an existence worthy of human dignity, and supplemented, if necessary, by other means of social protection.
4. Everyone has the right to form and to join trade unions for the protection of his interests.[3]

Social sustainability initiatives encourage companies to prioritize the rights and needs of workers, which can include living wages, safe working conditions, collective bargaining rights, equity, inclusion, and well-being.

These social goals seem obvious, indisputable—of course we expect people to be treated with respect, of course people need work to survive. And yet, companies continue to struggle to develop socially responsible workplaces. Social goals are difficult to implement and the success of any particular initiative or approach is often untested or uncertain. Workplace social responsibility requires a complex mix of disciplines—compliance, risk, behavioral science, psychology, diversity, and health—and a willingness to accept that progress is often slow and difficult to define.

As described in this chapter, lawyers play an important role in the advancement of social sustainability. Our laws recognize the significance of work in our communities and the importance of removing barriers to an individual's ability to earn a living and contribute to society through their work. Regulations in Canada establish accountability for the social aspects of work, such as equity, wellbeing, and inclusion. Our laws set the foundation for advancing social responsibility.

Employment lawyers[4] must know and consider these laws when advising clients—in other words, employment lawyers *are* social sustainability lawyers. And, as social expectations and working conditions continue to change, lawyers have an opportunity to influence the future of social sustainability initiatives.

Employment lawyers will play an important role in defining the line between voluntary social responsibility and mandatory legal obligations. As they work through challenging labour relations, disputes, and negotiations on behalf of clients, lawyers will influence how our laws need to adapt in the face of artificial intelligence, economic transition, flexible work arrangements, and contract-based work.

This chapter provides a brief history of Canadian workplace law and social sustainability, and then focuses on how lawyers continue to shape corporate social responsibility at the workplace.

[3] United Nations General Assembly. The Universal Declaration of Human Rights (UDHR). New York: United Nations General Assembly, 1948. Article 23. See: udhr.pdf (un.org).

[4] I use the term "employment lawyer" to include the legal disciplines related to the workplace, including employment law, labour law, and human rights laws.

2 Workplace Laws and Social Sustainability

2.1 A Primer on Canadian workplace laws

A brief overview of Canadian workplace laws is necessary to understand the connection between workplace law and social sustainability.

There are three main sources of Canadian workplace laws:

(a) The *Canadian Charter of Rights and Freedoms* protects procedural rights to strike and bargain collectively and imposes obligations on the government to protect the privacy and equality of workers.
(b) Provincial and federal legislation establish statutory rights and responsibilities, requiring employers to, among other things, prevent discrimination and harassment, create physically and mentally safe workplaces, narrow gender pay gaps, provide minimum wages, prevent excessive working hours, and enable workers to bargain collectively for their terms and conditions of work.
(c) The common law, developed through judicial and arbitral decisions, establishes procedural and substantive employment rights relating to duties of good faith, notice of termination, protection from mental distress and harassment, and privacy. Labour arbitration and employment litigation decisions are contextual and fact-specific, which results in an ever-evolving legal landscape in relation to the workplace.

Most workers in Canada have the right to seek representation by a bargaining agent if the majority of workers support unionization. In a unionized workplace, conditions of work are negotiated with the union. Collective agreements tend to have a three-to-five-year term. Decisions about workplace or policy disputes are made by arbitrators appointed by the union and employer.

In a non-unionized workplace, employers have greater discretion to change the conditions of the workplace. Remedies under Canadian laws tend to be complaint-based. Disputes are resolved through the Courts or administrative tribunals established under employment standards, human rights, privacy, and health and safety laws.

Historically, the legal framework in Canada was based on work as a commercial, transactional relationship—an exchange of "human capital" or "master and servant". Except for basic regulations relating to minimum employment standards, physical health and safety, and collective bargaining, companies were not responsible for the social impact of work, such as equity, inclusion, or well-being. Corporate social responsibility was not a significant part of an employment lawyer's day-to-day practice.

This has steadily changed over the past 40 years. Responsibility for workplace health, well-being, and equity—which are key parts of social sustainability—have become important considerations for lawyers when advising and advocating for their clients. Our regulations and the common law now recognize that work is an essential component of an individual's sense of identity, self-worth, and emotional well-being. Employment law is no longer merely about negotiating, bargaining or litigating the contractual parameters of a working relationship, but also determining the accountability and responsibility for the social aspects of work.

2.2 The "Master and Servant" Workplace

Until the 1980s, the legal system treated work predominantly as a transactional, contractual relationship between a company and worker, subject to the laws governing a "master and servant" relationship.[5] Regulations established basic rights to wages, safety, and collective bargaining rights, but most worker rights were set out in individual employment and collective agreements, which were matters of private contract to be determined by the parties and interpreted restrictively by courts and arbitrators.[6] The master and servant relationship was paternalistic, rooted in the command and control of an employer (the master) and the duty to obey of the worker (the servant).[7] The relationship was marked by economic dependence and the worker's vulnerability to management.[8]

This narrow conception of the workplace was consistent with thought leadership on corporate social responsibility at that time. In 1970, in a New York Times Magazine op-ed, Milton Friedman famously criticized the concept of corporate social responsibility to workers and reiterated his view that "there is only one social responsibility of business – to use its resources and engage in activities designed to increase its profits so long as it stays within the rules of the game, which is to say, engages in open and free competition without deception and fraud."[9] Friedman argued that managers were poorly prepared to make decisions about social justice, since their expertise is only in running a company—producing a product, selling it, or financing it.

From this perspective, work is a transaction, a straightforward exchange of supply and demand between a company and a worker. Worker rights focused on decent pay, hours, and security, not fulfilment and purpose.[10] This transactional view of work relieves companies and their shareholders from the ambiguity and uncertainty (and associated expense) required to manage the diverse perspectives and experiences of the workforce. Neither Friedman, nor the traditional legal framework, recognized the company's role in the broader social purpose of work, of enabling individuals to contribute to their communities, express their identity, and find a sense of belonging and self-worth.

The transactional view of work is quite far removed from the actual experience and expectations of many workers. While work is transactional, it also creates a

[5] See, for example, *Carter v. Bell & Sons (Canada) Ltd.*, [1936] OR 290 (C.A.).

[6] Archibald (2020), p. 6.

[7] See Stanley (1974), p. 395.

[8] *Ibid.* at 397.

[9] Friedman (1970).

[10] See also Tugend (2023).

community that—like our families, friends, neighbours, or political systems—has the power to inspire, irritate, or limit us. Our reflections on life, our choices, our sense of purpose and impact, apply to work as much as to other social institutions.[11] We spend most of our time at work. We want to work for organizations that are just, equitable and fair.

Academics have suggested that, before the 1970s, passion and purpose was not a priority for job seekers, and their focus was only on the transactional aspects of work (pay, hours, job security).[12] But arguably, the social aspects of work were not unimportant to workers; rather, the workers most impacted did not have the authority, voice, or ability to speak up about what was holding them back, such as racism, sexism, or exclusion of persons with disabilities.

It isn't necessary to buy into popular culture notions of "purpose-driven work" or "following your passion" to accept the idea that work is part of our social structure. Most people need to work to survive, and opportunities to have stable work are not available if workers are discriminated against, forced to work in unsafe conditions, or harassed at work. It is very difficult for workers to be productive when they face the burden of systemic racism, exploitation, mental distress or other psychological harms. These are the social issues that arise at work, and stand apart from the private, contractual negotiations or bargaining over wages, hours of work, position, or termination rights.

With increasing awareness of social impact from the 1970s to the present, it was inevitable that how we think about work, workplace laws, and social responsibility would shift from "master and servant" to something more equitable and more *human.* Equity, belonging, and purpose are now frequent topics of workplace discussion, and are more embedded in Canadian laws, legal systems, and workplaces.

2.3 Law and the Enablement of Workers

In the 1980s, there was a reframing of Canadian workplace laws to include this social perspective. A few notable examples are described below.

In 1981, Canadian labour lawyer and law professor Brian Langille wrote an essay to reframe the purpose of workplace laws. He observed that society wants to see that its citizens are treated with respect, as human beings capable of forming and acting upon their own ideas of how their lives should be lived. The purpose of the law was to establish standards to ensure economic survival and ensure that workers could contribute and express themselves as citizens.[13] He later expanded on this idea, identifying the purpose of the law to "identify and remove the obstacles to the

[11] Whether work "should" be a significant part of individual identity is debated. For a discussion of the "gospel of work", see Thompson (2019). See also Tugend (2023).

[12] An overview of this thought leadership is in Tugend (2023).

[13] Langille (1981).

deployment of human capital in ways which optimize its intrinsic and instrumental role in lives lived."[14]

Professor Langille's essay has been recently described as "path breaking" and as having a "lasting and positive impact on the way law teachers, legal scholars and some practitioners conceptualized the academic and practical discipline devoted to the understanding of how the Canadian legal system organizes and regulates activity at work".[15] His essay, and the thought leadership that followed, challenged the paternalistic, transactional approach to work, and encouraged lawyers to view the purpose of workplace laws as *enabling* people to find social purpose and economic stability through their work. To put this idea another way, employment lawyers should view laws as being aligned with the principles we now describe as social sustainability.

This approach to workplace laws was furthered in 1984, as provincial and federal governments considered the development of human rights and anti-discrimination legislation. Justice Rosalie Abella was appointed Chair of a Commission to examine equality and equity at the workplace, with a particular focus on the treatment of women and Indigenous people. The Abella Commission prepared a Report that is a foundational document in understanding the evolution of workplace equality rights in Canada. The Abella Commission recognized that one of the concerns to achieve equality was "those conditions in the workplace that militate against equal participation in employment."[16] Again, the legal framework of human rights law was premised on granting opportunities to individual workers.

The common law moved further from the "master and servant" framework in 1987, when the Supreme Court of Canada recognized the critical importance of work to individuals and communities. Chief Justice Dickson noted:

> Work is one of the most fundamental aspects in a person's life, providing the individual with a means of financial support and, as importantly, a contributory role in society. A person's employment is an essential component of his or her sense of identity, self-worth, and emotional well-being.[17]

Since the 1980s, there has been a proliferation of workplace legislation and decisions related to the social aspects of work, including from protection against psychological harassment at work, constitutional protection of bargaining rights, pay equity, workplace violence protection, accessibility, privacy, mental health, and modern slavery, to name just a few.

These Canadian legal developments were occurring at the same time as international human rights organizations were redefining concepts of sustainable development to include to social sustainability. In the 1987 report published by the World Commission on Environment and Development entitled "Our Common Future"

[14] Langille (2018).

[15] Archibald (2020), p. 2.

[16] Report of the Commission on Equality in Employment (Abella Report), Minister of Supply and Services Canada (October 1984) at 6.

[17] *Reference Re Public Service Employee Relations Act (Alberta)*, [1987] 1 S.C.R. 313 at p. 368.

(known as the Brundtland Report), the key message that actions taken by corporations should not compromise future generations, incorporated a triad model of sustainable development, that included ecology, economy, *and society*.[18]

In summary, the reframing of work in our legal system as "fundamental" and a means of "a contributory role in society" has an important connection to social sustainability, which focuses on whether communities can thrive, grow, and achieve economic stability. The purpose of workplace laws has evolved from focusing only on the transactional "master and servant" relationship to regulating the social aspects and impact of work. And, as a result, social sustainability has become a core part of practicing employment law.

3 Lawyers' Role in Social Sustainability

In all likelihood, only a handful of employment lawyers would describe their practice as focused on social sustainability. A significant part of practicing employment law still involves navigating traditional, transactional employment and labour issues—bargaining, contract negotiation, policy drafting, arbitrations, wrongful dismissals. Even when the practice involves compliance with social-related regulations, such as human rights, safety, wellness, or privacy, it is not generally framed as social sustainability, but as compliance with regulatory requirements. Social sustainability is more commonly considered to be an international law or policy issue, as opposed to a domestic employment lawyer's realm.[19]

Employment lawyers can create meaningful, impactful legal careers by reflecting more on the connection between the law and social sustainability. Employment lawyers play an important role in defining the line between voluntary social responsibility and mandatory legal obligations.

Also, as working conditions are changing at a rapid pace and as a result, employment lawyers have an opportunity consider and influence how future laws will enable workers in the context of artificial intelligence, economic transition, flexible work arrangements, and contract-based work.

3.1 Setting the Boundaries of Voluntary Corporate Social Responsibility

Canadian laws would play a lesser role in social sustainability if corporations voluntarily adopted social sustainability standards. For years (if not decades), workplace experts have criticized the reliance on legislation and compliance programs as inadequate and have urged companies to adopt proactive, preventative programs.[20]

[18] Brundtland Report, *supra* note 1.

[19] For a discussion of the role of domestic law versus foreign law, see Langille (2009).

[20] See, for example, Dobbin and Kalev (2016).

Many companies voluntarily recognize the need to create equitable, inclusive, and healthy workplaces. The UN Global Compact, signed by more than 20,000 companies worldwide, articulates the business case for social sustainability:

> … a lack of social development, including poverty, inequality, and weak rule of law, can hamper business operations and growth. At the same time, actions to achieve social sustainability may unlock new markets, help retain and attract business partners, or be the source for innovation for new product or service lines. Internal morale and employee engagement may rise, while productivity, risk management, and company-community conflict improve.[21]

Yet, companies face ongoing criticism of their commitment to social sustainability. What Milton Friedman perhaps got right was his skepticism that market forces alone would drive companies to create socially responsible workplaces. Even today, managers are poorly prepared to make decisions about the equity, safety, and wellbeing of employees.[22]

What prevents businesses from adopting socially responsible workplaces isn't necessarily an intention to cause harm to workers. Knowing where and how to allocate scarce resources in order to improve working conditions, equity, and wellness—especially when workers have individualized and often competing needs—is not a straightforward task. Focusing on shareholder value is more predictable, rational, and easier to measure than managing a complex workplace.

And so, relying on companies to adopt voluntary commitments is unlikely to result in consistent, sustained efforts to improve health and wellbeing, equity, and stability at work. Regulations and a legal framework are required to compel companies adopt socially responsible workplace practices.

Law and policy are not the entire solution—there are notable shortcomings about enforcement and accessibility to our complaint-driven legal system.[23] Nonetheless, mandatory legal requirements continue to play an important role in advancing social responsibility at work.

3.2 What Should Be Regulated?

Policymakers and businesses are continuously challenged by questions about the scope and impact of workplace social responsibility. Examples of these questions include:

- *Are companies only responsible for their own workers, or for any worker impacted by their business activities?* Does it matter how much control the company has over working conditions?
- *What standards of pay, hours of work, safety, labour rights, and human rights are expected when companies operate outside of Canada?* Are Canadian companies

[21] United Nations Global Compact. Social Sustainability | UN Global Compact.

[22] See for e.g. Mimbs Nyce (2022).

[23] Archibald (2020), p. 12.

operating abroad required only to comply with the laws in the jurisdiction in which they operate—or are Canadian legal standards the right baseline?

- *Who is responsible for the impacts of workforce transitions*? As companies transition to cleaner energy or adopt artificial intelligence and automation, who is responsible for the impact on workers and the changes (and potential reduction) in employment opportunities?
- *Are existing laws adequately enforced*? Administrative agencies are overwhelmed and resource constrained. An absence of effective enforcement mechanisms over human rights, pay transparency, accessibility, pay equity, and forced labour could undermine the social purpose of these laws.

There are no easy answers to these questions; our courts and regulators need to strike the right balance between assigning responsibility for workers and giving companies enough flexibility to determine what to prioritize and how to effectively enable workers in their industry, jurisdiction, and business model.

Companies, lawyers, and policymakers must continuously balance Friedman's concept of shareholder value with broader social sustainability goals. If laws are so complex or onerous that compliance requires extensive time and resources, companies may not be able to operate profitably, and will relocate, restructure, or close, which obviously has a significant impact on jobs and the economy. But, without any laws enforcing social responsibility, employees are at risk of being marginalized, underpaid, and exposed to unsafe working conditions, which also impacts the growth and stability of our communities.

Another challenge is that the practical application and effectiveness of social impact programs are difficult to assess. Determining the right approach to social sustainability lies not only in defining the boundaries between regulation and voluntary action (who is responsible, for what, and the consequences) but also in the practical application (how to operationalize social impact at the workplace). It is difficult to develop an empirical study of *what actually works* to improve working conditions, often because of a lack of data about workers and the difficulty of studying a policy or approach over a long period of time.[24]

Employment lawyers can contribute to this work in a number of ways. Critical thinking skills, an understanding of the legal framework, and experience in applying the law to a workplace are all important aspects of understanding the effectiveness of policy in regulating the social aspects of work. Also, if lawyers consider and share[25] outcomes from the legal process—not through the lens of whether they "lost" or "won" a case, but *why the dispute arose in the first place*—companies and workers would have more information to assess the effectiveness of operational programs relating to social sustainability. Did the approach taken by the employer contribute to or distract from the purpose of creating an equitable, safe workforce?

[24] See the discussion from Dobbin and Kalev (2016)—which describes what *doesn't* work.

[25] Subject to the caveat, of course, that any discussion of particular facts and outcomes is subject to solicitor-client privilege.

Were the worker's demands reasonable? What lessons can we learn from the effort and impact of compliance with a regulation?

In short, their skills, knowledge, and experience create opportunities for employment lawyers to play an important role in guiding people and organizations through challenging questions about social sustainability.

3.3 The Future of Workplace Laws

In addition to influencing the existing challenges of social sustainability, lawyers can play an important role in shaping the future of the workplace. Workplace conditions change constantly with the introduction of new technology, automation, remote work, outsourcing, and flexible work. Many of our workplace laws are based on a traditional view of work—long term, hierarchical, full time—and are not necessarily keeping up with changes to people, technology, and the global economy.

Lawyers entering the profession will have the opportunity to consider and shape how these laws could change. Examples of potential changes to our laws are described below.

The regulation of the "gig" economy is a recent subject of debate, disputes, and thought leadership, as project-based work becomes increasingly prevalent. Technology enables the creation of a talent marketplace that links workers to projects (the "gig" workers to their "gigs"). This changes the workplace structure from a traditional, role-based organization to a series of dynamic, project-based interactions. No more org charts, hierarchy, or long-term roles—the focus is on agility.

Flexible work can benefit workers—more choice, more agency, opportunity to work across industries and gain experience. But, flexibility also creates risks to workers. It is less certain whether laws designed to protect workers—such as human rights, mental health, pay equity, or compensation for job loss—apply to a "gig-based" work structure. Our laws are based on the worker's role, status, classification, and relationship to the company.

Another debate relates to responsibility for outsourcing. When companies outsource manufacturing or other aspects of their business, whether to overseas companies, labour agencies, or franchisees, they retain the financial benefit of the business but offload workplace risks and costs to suppliers.[26] There are existing laws that enable a Court or arbitrator to allocate business risk for some outsourced workers—such as joint or related employer requirements—but these laws require a degree of integration and control between the company and the workers.[27] Factors such as common ownership, financial control, control over labour relations, common management, and interrelationship or interdependence of operations and

[26] For a discussion of "offloading" in American manufacturing, see Unseem (2024).

[27] *Modern Cleaning Concept v. Comité paritaire de l'entretien d'édifices publics de la région de Québec*, 2019 SCC 28 at 34.

representation to the public as a single integrated business[28] do not apply across a complex, extended supply chain. Only recently, the Canadian government has introduced the idea of accountability for workers in the supply chain, with the enactment of the *Fighting Against Forced Labour and Child Labour in Supply Chains Act.*[29]

When the relationship between the workers and the company is periodic, short term, or indirect, allocation of social responsibility is less clear. Labour rights organizations, both in Canada and the United States, have argued that broader based bargaining is necessary to ensure workers can engage in meaningful collective bargaining and are protected by labour rights statutes.[30] To date, these suggested changes to the law have not been adopted.

Dalhousie law professor Bruce Archibald recently wrote that "underlying economic and social conditions have largely rendered traditional labour and employment law to be impediments, rather than supports, in the attainment of social and economic justice for the vast majority of Canada's working people."[31] Professor Archibald's suggestion is to change our laws to assign accountability for workers to the person or entity who benefits from the efforts of the worker and influences the nature and conduct of the work.

How much do our laws need to change to protect workers? Are *more* laws the answer, or do we need a wholesale change to the legal framework? The future of our workplace laws is complicated, unpredictable, and messy – and present an opportunity for sustainability-minded lawyers to shape and have an impact on how workplace laws should evolve to continue to promote "the expansion of real capability to lead lives we have reason to value."[32]

4 A Personal Journey Into Social Sustainability

In my own career journey, I became an employment lawyer because of my passion for people and the workplace. It has led me in many different directions, consistently focused on building sustainable workplaces and communities.

I am an employment lawyer and a business transformation consultant. I was an executive-level leader in talent, human resources, and legal operations. I was a partner at a national law firm, a litigation boutique, and a global law firm. I am now a partner at a global professional services firm, where I get to use what I have learned

[28] See for e.g. *Concerned Contractors Action Group*, BCLRB No. 32/86, (1986), 13 CLRBR (NS) 121 at para 166; see also *Sobeys Capital Incorporated v United Food And Commercial Workers International Union*, Local 1518, 2020 BCLRB 97.

[29] S.C. 2003, c. 9.

[30] See, for example, The Re-emergence of Sectoral Bargaining in the US, Britain, Australia and Canada | OnLabor (accessed September 6, 2022). Also, see Archived - The Changing Workplaces Review – Summary Report | ontario.ca (2016). The *Changing Workplaces Review* recommended amending Ontario employment and labour laws to expand franchisor liability and create a certification model for the franchise industry. These proposed changes were ultimately not adopted.

[31] Archibald (2020), p. 9.

[32] Archibald (2020), p. 27 quoting Brian Langille.

in operational leadership and law to transform organizations—how they work, their culture, their technology, how they navigate change and uncertainty. My story is less about career planning and mapping, and more about exploring what I found interesting and impactful.

Before going to law school, I was passionate about equality, opportunity, and human rights, but thought that working in human rights law meant moving to the Hague, becoming an international lawyer, working alongside governments or NGOs to tackle global human rights issues globally—across the world, not close to home. And that was not the vision of my future. I wanted to work in Canada, in my community, and make an impact on the people I saw every day.

I saw a glimpse of the future in my first month at Dalhousie law school. A union-side and management-side labour lawyer presented to our first-year class about a Canadian pay equity case. The presentation was about the legal process—its complexities, the procedural steps—but I was struck by what was at stake for Canadian workplaces and, by extension, for my friends, family, and community. At that time, the concept of pay equity[33] was still in the early stages of being defined and developed. Gender predominance? Male and female roles? Job evaluations? Equity? It opened my eyes to the complexity and very real need to create systems and rules that ensured Canadian workers had equal opportunity to earn a living. And, by becoming an employment lawyer, I could do that work close to home.

During my time at Dalhousie, I was very fortunate and grateful to learn from two brilliant, insightful, and passionate labour law professors, Innis Christie and Dianne Pothier, who encouraged me to think about the links between human rights, equity, work, and labour relations. They both contributed greatly to Canadian laws and policy that enable workers, promote equitable systems, and define social responsibility.

I started my career as a management-side employment lawyer at a national law firm that was well known for its workplace practice. I had a chance to work with exceptional lawyers on cases that required critical thinking about workplace laws. 11 years later, the firm collapsed—which was devastating, and also taught me about the importance of leadership and talent management. I learned that a positive workplace culture doesn't just magically happen, but is part of intentional strategy, planning, and execution of people-focused systems, policies, and tools.

In 2014, I joined an international law firm, as a partner in the labour and employment practice and, three years later, had the opportunity to become the firm's Chief Talent and Practice Officer. I learned about how teams work together to build a culture, the importance of having engaged, inspiring leaders who are willing to challenge the status quo, and the conditions we all need to thrive at work.

Since 2022, I have practiced at a global professional services firm, and continue to learn and build on the intersection between people, law, and operations. Working alongside professionals across many disciplines—law, accounting, finance, risk management, environmental, human resource and technology consultants—enables

[33] Pay equity is equal pay for work of equal value. The concept is now embedded in legislation in Ontario, Quebec, and for federally regulated organizations.

our team to support organizations more holistically. I have learned that we work better, together—as our CEO says, in "supercollaboration"—to solve complex workplace issues and create more sustainable workplaces and communities.

I try to use everything I have learned in the past 20 years from my professors, law colleagues, and business professionals to enable strong, collaborative teams and advise organizations on navigating laws regulating social responsibilities at workplace. Becoming an employment lawyer was the foundation for a career in social sustainability.

5 Conclusion: How Lawyers Can Lead on Social Sustainability

Creating a socially responsible workplace requires a complex mix of disciplines—compliance, risk, behavioral science, psychology, diversity, and health, to name just a few. The competing and confusing voices and opinions sometimes make it nearly impossible for workers to understand their rights, or for companies to know where to start and how to move forward.

Lawyers have an important role to play in enabling social sustainability at work. Employment lawyers have the experience, the opportunity, and arguably, the responsibility, to make the laws known, clear, and more easily complied with.

Also, the complexity and impact is what makes this area of the law rewarding. A legal practice focused on social sustainability is about more than achieving a desired goal or outcome in a particular case. The reward is in continuously contributing to building a community where businesses and workers can grow and thrive, belonging exists, accountability and responsibility is known and appreciated.

References

Archibald BP (2020) Labour law as a subset of employment law? Updating Langille's insights with a capabilities approach. Dalhousie Law J 43:2

Dobbin F, Kalev A (2016) Why diversity programs fail. Harv Bus Rev, July-August 2016

Friedman M (1970) A Friedman doctrine – the social responsibility of business is to increase its profits. The New York Times of 13 September 1970 (nytimes.com)

Langille B (1981) Labour law is a subset of employment law. UTJL 31:200

Langille B (2009) What is international labour law for? Law Ethics Human Rights 3(1):46

Langille B (2018) Human development: a way out of labour law's fly bottle. In: Collins H, Lester G, Mantouvalou V (eds) Philosophical foundations of labour law. Oxford University Press, Oxford

Mimbs Nyce C (2022) What Managers Can Do About Burnout. The Atlantic, November 5, 2022.

Stanley DC (1974) Prerogative in private and public employment. McGill Law J 20(3):394

Thompson D (2019) Workism is making Americans miserable. The Atlantic, February 24, 2019

Tugend A (2023) Is following your work passion overrated. The New York Times, August 3, 2023

Unseem J (2024) Boeing and the dark age of American manufacturing. The Atlantic, April 20, 2024

Jillian Frank joined KPMG as a Partner in 2022. Over the past 20 years, she has practiced labour, employment and human rights law for national and global law firms. Jillian regularly advises clients on social sustainability and human rights, including recently implemented forced labour reporting obligations in Canada. She is the National Lead of Legal Transformation, Technology, and Managed Services at KPMG Canada.

Jillian is also a Chartered Professional in Human Resources and draws on her past operational leadership experience as a CHRO to provide practical advice and services to clients. Her experience in leading a human resource function included delivering best practices in employee and stakeholder engagement, promoting safety, equity and inclusion at the workplace.

Beyond the Bar: Purpose, Risk and Innovation

David G. Myrol

Abstract

This chapter traces the professional and personal journey of David G. Myrol, K.C., a lawyer who built a successful career in Occupational Health and Safety (OHS) law, eventually becoming a senior partner at a top-tier law firm before taking the bold step of leaving it all behind to pursue entrepreneurial ventures. The narrative highlights the early influences of a philosophy professor, who encouraged the author to push intellectual boundaries, eventually leading to a law career shaped by resilience, adaptability, and a commitment to lifelong learning. After facing early career setbacks during the 1990s recession, the author found an unexpected path in OHS law, embracing a niche area of prosecution that would later define his professional life.

The chapter also explores a pivotal moment in an OHS case that underscored the importance of workplace safety, driving the author's dedication to this field. Serendipity played a role as well, with career shifts, such as being recruited by McLennan Ross LLP, leading to the development of one of Canada's leading OHS law practices. In addition to legal accomplishments, the author pursued further education, earning an MBA and forming successful business ventures.

Concluding with reflections on purpose and the importance of family support, the chapter underscores the themes of taking risks, embracing new challenges, and committing to continuous growth. Even with a storied career behind him, the author views his next chapter with enthusiasm, demonstrating that success is as much about finding purpose as it is about achieving professional milestones.

D. G. Myrol (✉)
National Safety Services Inc., Edmonton, AB, Canada
e-mail: david.myrol@nationalsafety.ca

A. McLaughlin, C. Chell (eds.), *Building the Case: How Lawyers are Leading on Sustainability*, https://doi.org/10.1007/978-3-031-88455-9_12

1 Introduction

I built a successful career as a leading expert in Occupational Health and Safety (OHS) law. As a senior partner in a prestigious law firm with a client list that any firm would envy, I had reached a pinnacle of professional success. The path ahead was clear and comfortable. But I was unsettled and woke up one day and a made a decision that defied conventional wisdom—I walked away from it all. I left behind the security, prestige, and my identity for the last 30 years. I made the decision to pursue my business interests, including taking on the role of a CEO at a startup technology company focussed on workplace safety. The company was undercapitalized and entering a technology market spinning at a dizzying rate of change from advances in artificial intelligence. With a 95% failure rate, it was not exactly a recipe for success.

To most, my decision seemed reckless, even irrational. Why risk everything I had worked so hard to build? The truth is, looking back, all my career decisions were infused with an entrepreneurial spirit, a desire to go where no one else has gone, to build something new, something meaningful. I also knew deep down, at a molecular level, that I would never be happy unless I took this path, wherever it might take me, for better or worse. While most would question my sanity, I was at peace with the decision.

While I'm hesitant to offer advice, there is one truth I have come to know: clarity comes with purpose. Even the boldest decisions become easier when driven by a clear sense of purpose. I can't help you find your purpose, but I can share how I found mine. This is my story.

2 The Formative Years

Many of us have a mentor or teacher who shapes the trajectory of our lives, and for me, that person was Dr. Wes Cooper, a philosophy professor at the University of Alberta. At the time, I was completing an arts degree and thinking about law school. In my final year, I enrolled in Dr. Cooper's philosophy of law class—not out of any particular passion for philosophy, but because it fit my schedule and seemed relevant to my plans. Dr. Cooper saw something in me that I hadn't yet recognized in myself. I wasn't the best philosophy student, and I lacked confidence, often feeling intimidated by my articulate classmates. Yet, despite my insecurities, Dr. Cooper encouraged me to pursue an honors degree in philosophy—a challenge I hadn't previously considered.

Flattered and curious, I committed to an additional year of intense study, taking a full course load of demanding philosophy classes from Aristotle to Nietzsche. It was a rigorous year that sharpened my critical thinking skills and stretched my intellectual boundaries. Ultimately, it also helped me realize that law, with its immediate and lasting impact on real life, was the right path for me. After completing the degree, I applied to law school, was accepted, and never looked back.

To Dr. Cooper, wherever you are, your influence on the direction of my life was profound, and I am forever grateful.

Law school was a pivotal period in my life in more ways than one. It's where I met my wife, Denise—now Madam Justice Denise Kiss—whose partnership has been one of my greatest blessings. Together, we have two wonderful children, Lucy and Carter, who bring us constant pride and joy. Beyond family, law school gave me lifelong friendships—friends who not only helped me through those challenging years but also enriched my life in ways I couldn't have foreseen. Looking back, law school wasn't just about learning the law; it was about surrounding myself with good people who shared my values. It was about building relationships with those who make you better.

3 Early Career Challenges

My early years in law were anything but smooth. The recession of the early 1990s left many of us, myself included, out of work with limited opportunities. At the time, it felt like a major setback, but in hindsight, it made me more resilient and sharpened my skills. During that period of uncertainty, I gravitated toward criminal law, which led me to accept a one-year contract with the Crown Prosecutor's Office. I initially saw it as a stepping stone toward becoming a stronger criminal defence lawyer. What I didn't anticipate was how much I would enjoy the work and the people. That one-year contract turned into nine years, and it was during that time that I found my way into Occupational Health and Safety (OHS) law—a field that would define my career.

4 Developing Purpose and Expertise

At the Edmonton General Prosecutions Office, we carried the heaviest caseload in the country. We were overworked and underfunded, but I thrived in that demanding environment. Being in court every day pushed me to grow, think on my feet, and adapt quickly. I loved nearly every minute of it. Then one day, a memo arrived from our Chief Crown—this was before email existed—seeking volunteers to specialize in niche areas of prosecution. The goal was to reduce our workload through specialization. One of the areas listed was OHS law.

At the time, I knew nothing about OHS, and I wasn't particularly interested in it. But I respected my Chief Crown and wanted to contribute, so I volunteered. Little did I know, that decision would shape the rest of my career. What began as a routine assignment quickly evolved into something far more significant. I led a pilot project as Alberta's first dedicated OHS Prosecutor—a model that would later be adopted by other provinces. I also developed training programs for OHS Officers to improve their ability to investigate workplace incidents. Eventually, I left government to become Canada's first full-time OHS defense lawyer.

That decision—to step outside my comfort zone and take on something unfamiliar—became the pivotal moment in my professional life. It taught me that embracing uncertainty with curiosity and commitment often opens doors to opportunities we couldn't have anticipated.

5 A Defining Moment

I remember the exact moment I knew OHS law would be my specialization. I was preparing one of my first OHS cases, which involved a young man who had been badly burned in an oil and gas explosion. I met him at the University of Alberta Hospital's burn unit, where he walked me through the details of his accident. At the end of the interview, he thanked me for listening and tried to stand up to shake my hand. He couldn't. He apologized and explained that so many skin grafts had been taken from the back of his legs that his legs could no longer straighten. The simple act of standing—something most of us take for granted each and everyday - was now impossible for him.

His pain and suffering were unnecessary and preventable, and in that moment, I realized the profound importance of workplace safety. It was then that I decided to dedicate my career to OHS law, and I've never looked back. I've never let anyone deter me from that purpose, and I've never taken for granted the simple act of standing up.

6 Serendipity

Most successful people will admit that luck plays some role in their journey. I wish I could say I predicted the rise of OHS law in Canada, or that I knew the Alberta government would intensify its efforts to keep workers safe, or that Canada would amend its Criminal Code to hold organizations more accountable for workplace negligence. All of this change was happening as I was building my career in OHS law, and I was fortunate enough to catch that wave.

Another stroke of luck—though I didn't see it that way at the time—was when my new Chief Crown decided to move me out of OHS prosecutions into a serious crimes unit. I disagreed with the decision, but I had a choice: accept the move or leave. That's when I was recruited by McLennan Ross LLP, a firm with the vision and commitment to support me in building an OHS practice group. I had the benefit of working alongside a group of highly skilled lawyers and exceptional support staff.

I spent the next 20 years with McLennan Ross, where we became one of Canada's leading OHS law firms. I practiced OHS law across Western Canada, became a member of four different law societies, and had the privilege of writing, presenting, and shaping the development of OHS law through my work, including several landmark decisions.

I'm leaving OHS law in good hands, with a group of talented colleagues who share my deep commitment and passion for this critical area of practice. I look forward to following their careers, and watching how our courts will continue to interpret and shape the future of this important domain. My career in OHS law has been immensely rewarding, and as I transition forward, I do so knowing that the field is in the capable hands of those who will continue to drive its progress.

7 Lifelong Learner

In 2016, I obtained my MBA through the Haskayne School of Business, while carrying a full practice, and with teenage children at home. Those years are largely a blur. I recall flying home from Grande Prairie after an intense week of trial. I got on the plane, exhausted, flipped open my text on statistics and started studying for my exam the next morning. It was a crazy, grueling time, and I loved it. In my MBA program I met my business partners, Karl Michl and Doug Craig, and together we started a private investment company which is now thriving and has already exceeded our wildest dreams. It's amazing what you can accomplish when you dream big and commit to the work. I'm proud to be their business partner, and proud of what we accomplished, including the jobs we have created, together.

In 2020, I obtained my ICD.D designation from Rothman's School of Business, and I continue with my lifelong commitment to learn, which includes new languages, venture capital and artificial intelligence. If anything has separated me from the crowd, it's been my work ethic and my passion to learn. I hope to carry those aspects to my last days.

8 Conclusion

Although I've achieved a great deal in my legal career, I believe the most exciting chapter is just beginning. While many of my peers are preparing for retirement, I feel more energized than ever. I have no plans to retire—there's still so much I want to accomplish.

None of this would have been possible without the support of those around me. I owe deep gratitude to my parents, whose work ethic and strong moral compass shaped who I am today. Their love and guidance laid the foundation for everything I've achieved. I've also been incredibly fortunate to have a loving and supportive spouse, who has been the cornerstone of our family. Her unwavering strength made it possible for me to pursue my ambitions, and for that, I am eternally grateful.

David G. Myrol K.C., ICD.D, MBA, LL.B., B.A. (Hons.), is a nationally recognized leader in Occupational Health and Safety (OHS) law and a key figure in shaping its development in Canada. As a former Crown Prosecutor and then senior partner at McLennan Ross LLP, David played a pivotal role in numerous landmark OHS cases, earning widespread respect for his legal expertise and extensive trial and appellate experience. He was Alberta's first designated OHS Crown Prosecutor and later became Canada's first lawyer to practice exclusively on OHS law, influencing significant legal precedents and regulatory frameworks.

In addition to his legal career, David transitioned into the business world, co-founding a private investment company and co-founding and leading National Safety Services Inc. as CEO, with a mission to revolutionize workplace health and safety. His entrepreneurial drive reflects his commitment to innovating and improving safety practices across industries. A prolific writer and speaker, David continues to advise government and industry, remaining at the forefront of workplace safety while pursuing new business ventures aimed at driving positive change.

Dissecting Sustainability: A Critical Approach to the Sustainable Development Discourse

Carlos Escoto Carranza

Abstract

This chapter critically examines the concept of sustainability, arguing that it has become a hegemonic discourse that perpetuates economic growth while failing to address genuine ecological and social issues. The author, Carlos Escoto Carranza, reflects on his experiences as an environmental lawyer in Mexico, tracing the historical adoption of sustainability from the 1972 "Limits to Growth" report to the Brundtland Report in 1987. He critiques sustainability through a Gramscian lens, suggesting it actually serves economic interests and calls for a re-evaluation of the concept to foster meaningful environmental and social change.

1 Introduction

When I was approached by the editors of this engaging and thought-provoking book, the premise for my chapter was to write a small account of my experience in the sustainability sector, talking about the challenges I have faced in my practice as an environmental and ESG lawyer in Mexico. As I reflected profoundly on this subject, deciding what would be a suitable narrative that would make my contribution interesting and appealing, I realized that sustainability has become a sort of monolithic concept, one that defines and frames our way of thinking, acting and practicing the law and pursuing development, particularly in a country with such contrasts and inequalities as mine.

So instead of writing about all the undeniable benefits and improvements that the sustainability framework has brought into our world (a topic that, in addition, has

C. E. Carranza (✉)
Galicia Abogados, S.C., Mexico City, Mexico
e-mail: cescoto@galicia.com.mx

A. McLaughlin, C. Chell (eds.), *Building the Case: How Lawyers are Leading on Sustainability*, https://doi.org/10.1007/978-3-031-88455-9_13

been thoroughly addressed in available literature and even other contributions along this book), I opted to follow a completely different course.

I thought that, perhaps, my contribution would be more valuable if this very concept - the one that I have devoted all my life to- was questioned, challenged, and put through the acid test of criticism, in the hope of moving the reader to question his own assumptions and preconceived notions on this subject and push the discussion forward.

In this sense, if this chapter at the very least irks you, the reader, makes you shake your head in contempt and disapproval, or elicits any other kind of intellectual counterarguments or response, then I am satisfied, for I have reached my main objective, which is to make you engage with me and think.

To be sure, in the face of the climate and environmental crises that are already upon us, I believe it is crucial to start thinking beyond our current models -which are running on fumes-, and to generate discussion, consider extraordinary approaches to solving our problems and, beyond all, foster an inquisitive and critical mind, for the thinkers are the ones that keep humanity moving forward.

2 Setting the Scene

As obvious as it may seem, sustainability is not an object: something that exists in nature which can be perceived through our senses or deducted from observation, and upon which objective conclusions and consent can be reached. To the contrary, sustainability is, in the first place, a concept, and a somewhat new one at that. Indeed, as we all know, the concept of sustainable development was coined almost 40 years ago in the Brundtland Report as the "*development that meets the needs of the present, without compromising the ability of future generations to meet their own needs*" (WCED 1987, p. 8).

This little phrase started a seismic shift in the collective west's worldview. Indeed, who would not want inter-generational justice? Who would not want to preserve our planet, in our never-ending pursuit of the next shiny thing?

Sustainability gave the perfect answer: economic growth today, indefinitely maintained and projected into the future. It is hard to contest the ideal that we should ensure the economic wellbeing of the humans that exist today, without forgetting the humans that will come tomorrow.

And so, the sustainability ball started rolling 40 years ago and has since become a common hand term in our western economic-driven societies, it is a usual buzzword in politician's speeches, corporate salespoints, product placement, universities' syllabi, fashion and even mass media entertainment.

As many young lawyers (and old ones too) know very well, choosing an area of specialization is the second most daunting decision one can take (the first one being choosing law school itself - at the time, marriage was still a nebulous and unappetizing concept that thirty something year old me would have to deal with).

Fresh off my second year of law school, I was also faced with this Sword-of-Damocles-like decision. My friends and classmates were all already working, clerking in big M&A or commercial litigation firms - boasting about their high salaries and the huge deals they were in.

Being a naïve and idealistic 21-year-old, I chose to become an environmental lawyer (without having the slightest clue of what environmental law was), when I started clerking in the environmental practice of a small firm - trying to claim my place in the world and hoping to make it a little better.

Working as a lawyer in a somewhat novel practice area (as opposed to the traditional civil or commercial practices that have been around forever), in a developing country, proved to be a real eye-opener (to say the least): on the one hand, I had to understand the scientific basis of the laws and regulations that set forth allowable thresholds of pollutants in, say, wastewater discharges, and on the other, I also had to engage with authorities that at times only used the law as a quick way to meet their inspection quotas or, even worse, make some extra money to take back home.

However, it was also encouraging to witness how companies - regardless of the imperfect regulatory framework - were committed to understand the baseline of their regulatory obligations and make the necessary investments to comply and go beyond the mere letter of a standard or the law.

Indeed, that has always been the most exciting part of my work as a sustainability lawyer, holding interdisciplinary discussions with academics and investigators at universities, regulators at public agencies or the EHS technicians at a plant, to try to frame a complex and dynamic subject matter as is the environment and our relationship to it, into something as rigid as the law, *and then put it into action!*

This is something that a contract or a torts lawyer will never understand: the mind-expanding effect of different disciplines engaging and working together, to outline positive feedback loops of environmental performance in our economic models (albeit with all the caveats that I will delve into in this chapter).

So it was that, when I started clerking, the sustainable development discourse was already well established as a firm (and inescapable) model for progress, and I fell for it hook, line and sinker. As I progressed in my career and developed my understanding of the intricacies of my specialization, such as the prospective and preventive nature of the environmental impact assessment, the importance of environmental baselines in assigning site remediation liabilities, or defining additionality in carbon projects, I thought I understood what sustainability was and how environmental law, as a regulator of human conduct, contributed to it.

As the saying goes, the grind catches up to all of us, and I realized that something was (and is still) amiss. Something brewed in the back of my mind, itching like the proverbial neck of a wool sweater.

How come, 20, 30, 40 years into the sustainability movement and alleged economic development, were deforestation rates still on the rise[1]? How come climate change was not abated, but greenhouse gas emissions were actually increasing[2]?

[1] https://www.globalforestwatch.org/dashboards/global/?lang=en.

[2] https://wmo.int/news/media-centre/greenhouse-gas-concentrations-hit-record-high-again.

How come more species are becoming extinct,[3] rivers are more contaminated[4] and plastic bags are being found in the bottom of the Mariana Trench[5]?

Somehow, I felt that we were all being taken for a ride and that, even in all its goodness and promise, sustainability was not working, and I had to understand why.

3 A Super-Condensed History of the Adoption of Sustainability

In the span of 40 years (a rather short timeframe, when viewed in the context of historical movements and ideas), sustainability has become a ubiquitous term, that permeates - at least at the discourse level- to most aspects of our lives, from the way we shop, do business, dress, eat, teach and *think*, to the point that it is no longer questioned or even thought about (albeit this is something that will be examined further ahead in this chapter). Indeed, it would not be an exaggeration to say that sustainable development has become almost as integral to the current articulation of western culture, as capitalism, democracy or freedom.

It may thus be surprising (particularly to younger practitioners) that sustainability is a concept that far from evolving as part of a culture in an organic way, it was "adopted" at a specific and crucial point in our recent time.

In 1972, the Club of Rome[6] commissioned the Massachusetts Institute of Technology to assess the link between our patterns of economic and population growth and consumption, in relation to the ecosystems that make possible this development and provide for our sustenance.

The resulting work was the seminal report Limits to Growth (Meadows et al. 1972), which concluded after various computer models were run, that economic and population growth interact with each other in a positive feedback loop, where in a world with finite resources, the stress in ecosystemic carrying capacity will check both the economic and population systems, eventually ending in a systemic collapse of both, due to natural resource depletion[7]:

> The industrial capital stock grows to a level that requires an enormous input of resources. In the very process of that growth it depletes a large fraction of the resource reserves available. As resource prices rise and mines are depleted, more and more capital must be used for obtaining resources, leaving less to be invested for future growth. Finally, investment cannot keep up with depreciation and the industrial base collapses, taking with it the service and agricultural systems, which have become dependent on industrial inputs. [...]

---

[3] https://wwf.panda.org/discover/our_focus/biodiversity/biodiversity/.

[4] https://www.globalcitizen.org/es/content/clean-dirty-rivers-water-polluted/.

[5] https://education.nationalgeographic.org/resource/plastic-bag-found-bottom-worlds-deepest-ocean-trench/.

[6] The Club of Rome is a think tank created in 1968, "*to address the multiple crises facing humanity and the planet. Drawing on the unique collective know-how* [of its] *100 members* [...] *we seek to define comprehensive solutions to the complex, interconnected challenges of our world*".

[7] Meadows et al. (1972), p. 125.

> Population finally decreases when the death rate is driven upward by lack of food and health services.

This conclusion, coming after a study commissioned by some of the world's most powerful industrialists nonetheless, sent shockwaves through the status quo, as it directly challenged the core notion of western economic models: that economic growth as measured by increase in the gross domestic product, cannot be sustained indefinitely, and if we continue with our "*present way of doing things (...) the basic behaviour mode of the world system is exponential growth of population and capital, followed by collapse*".[8]

The message underlying the conclusions reached by Limits to Growth, is that the economic process should be framed in the context of the environment that makes it possible, so that our resources are not depleted. In other words, economic growth should be curtailed and aimed towards a more static framework, aiming to rather balance the distribution of the resulting wealth that is created through the economic process.

The pushback and opposition to the conclusions reached by Limits to Growth were swift, clear and unequivocal in both the developed and developing world.

On the one hand, critics in the geopolitical North would not even entertain the idea of halting or even re-directing economic growth, as this would attempt against the very fabric of social development and progress (Sandbach 1978). As the Chairman of the Board of Directors of General Motors would vehemently state at the time[9]:

> (...) The consumer is the loser when irresponsible criticism and ill-conceived legislation break down faith in our economic system, (...) when the very idea of free enterprise is diminished in the eyes of young people who must one day manage our businesses. The dull cloud of pessimism and distrust which some have cast over free enterprise is impairing the ability of business to meet its basic economic responsibilities - not to mention its capacity to take on newer ones.

On the other hand, in the developing world, Limits to Growth was also received with much skepticism, as it implicitly suggested that the wide gap between wealthy and poor nations would never be bridged, as our planet would not endure billions of former poor people living in the same affluence as the population of industrialized countries (Dryzek 2005).

In the wider cold war era context, the geopolitical South perceived that Limits to Growth was a ploy by the First World to hinder its legitimate aspirations for development and keep it as a mere provider of natural resources that would fuel the North's development and progress (Biel 2007).

Thus, developing nations were swift in also rejecting the conclusions and suggestions posited by Limits to Growth, asserting their right to pursue economic

[8] Idem, pp. 128, 142.

[9] *Cfr.* Sandbach (1978), pp. 511–512.

growth in the terms and direction that each nation deemed fit, as this was an inescapable condition to escape poverty.

However, the study set in motion a series of worldwide discussions, summits and controversies that culminated in the publication of the Brundtland report in 1987, and the worldwide adoption of sustainable development as a paradigm.

4 Sustainability as a Discourse

The Brundtland Report sets forth the most accepted formulation of what sustainability means: development "*that meets the need of the present without compromising the ability of future generations to meet their own needs*" (WCED 1987).[10] At its core, this concept is sufficiently ambiguous to mean everything and nothing at the same time, encompassing promise and wellbeing for everyone, everywhere, now and forever, while committing nothing to achieve this goal, which makes such an appealing concept.

But this handy and attractive term merits closer inspection. I believe that sustainability has morphed from being just a proposition and/or expression of goodwill, to becoming a complex and confusing discourse, that is to say, "*a shared say of apprehending the world, that enables those who subscribe it to interpret bits of information and put them together into coherent stories or accounts*".[11]

In this sense, the sustainability discourse underpins economic, environmental, political and sociological elements that are not entirely innocuous or exempt from being loaded with the interests and aspirations of the parties that are articulating the discourse. Once the discourse has been formulated, it is then accepted and espoused by its adherents, who may not entirely be aware of the exercise in power that lies beneath the concept.

To be sure, discourses are embedded with power (Dryzek 2005), and an analysis of them "*reveal the power relations between competing actors occupying a common action space*".[12] Power, in this context, should not necessarily be understood in its direct, physical manifestation (i.e. as a show of coercive force that compels the recipient to act in a certain way), but rather, as succeeding in making a third party to behave in a particular manner.

In this framework, power is to be recognized as the "*ability of the powerful to affect the actions and thought of the powerless*",[13] which becomes more insidious if it is exercised through the regulation of the recipient's ideas and way of thinking, imposing certain interests as contributing to the 'common good' (Schmink and Wood 1987).

In the condensed storyline of sustainability, the discourse began with the implicit recognition made in Limits to Growth that the development course undertaken by

[10] WECD (1987), p. 8.

[11] Dryzek (2005), p. 9.

[12] Skillington (1997), p. 5.

[13] Gaventa (2006), p. 24.

the geopolitical North, was excluded to the rest of the world as our planet's biosphere would not resist the toll. On the other hand, economic growth was perceived as absolutely necessary to improve the livelihood and wellbeing of the masses of people in the global South (Dryzek 2005).

It was in this sense that the sustainable development discourse was born out of a sense of conflict between the developed and developing worlds, where the former was advocating for restraint in order to protect the environment, and the latter saw this effort as a move to hinder its legitimate development and economic growth rights, rejecting in consequence any environmental conservation efforts that would stop them from overcoming poverty (Peet and Watts 1996).

With the publication of the Brundtland Report and its swift adoption worldwide, a perceived breakthrough in these seemingly two opposing views was achieved, as it caters to the interests and concerns of all parties involved (Dresner 2008).

Thus, the sustainability discourse - a shared way of apprehending the word - was formulated and swiftly adopted the world over, which had the effect of preserving a basic tenet of the economic model prevalent for the last 150 years: economic growth must not be limited. Economic growth must go on.

Even today, 50 years after Limits to Growth, the notion of curtailing economic growth seems counterintuitive, even politically unthinkable (Daly 2007) with sustainability being seen as the antidote that will make perpetual growth possible, even in a world with limited resources.

The success of the sustainability discourse is evident and total. The principle of sustainable development has been enshrined in the political constitutions of most countries in the world -particularly in Latin America - and is perceived as the only available model for economic development and generation of wealth and wellbeing.

However, with the deforestation, climate change, species extinction and even population rates increasing as mentioned in the first Section of this chapter (which paradoxically confirm the findings of Limits to Growth), the hard questions must be asked: is sustainability effective in eliminating the consequences of economic growth? Has sustainability improved the state of the world? Has it not masked, rather, the symptoms of the illness and made it worse? More importantly, who is benefitting from maintaining the status quo?

5 Sustainability as Hegemony.[14]

Antonio Gramsci (1891–1937) was an Italian Marxist philosopher, who was co-founder of the Italian Communist Party, just before World War II erupted. The fascist regime in Italy imprisoned him for his political leanings, and while in jail, Gramsci collected his views on philosophy, economics, politics and sociology in journals that later became a required read for political science students and analysts.

[14] For this section, I have drawn from the analysis that I delved into in my Master degree's dissertation. *Cfr.* Escoto (2009).

In these "Prison Notebooks" (Gramsci 2007), Gramsci matured and articulated the concept of hegemony, which he understood as "*the 'spontaneous' consent given by the great masses of the population to the general direction imposed on social life by the dominant fundamental group*".[15]

Gramsci's main argument is that strong elites do not require coercion to guide the behaviour of the masses, but rule almost entirely through hegemony, reserving violence to the isolated geographical or social trouble spots that prove stubborn (Adamson 1980).

It becomes evident that hegemony is based on consent, which is almost always granted unconsciously by the lower strata in society. Indeed, hegemony is consent forged in the cultural, intellectual and moral realm – what Gramsci calls the "superstructure" (Birchfield 1999)-, where the values of the elites become the 'common sense' and 'natural' way to see the world (Rist 2007). To further drive the point, Gramsci argues that common sense is "*the conception of the world which is uncritically absorbed by the various social and cultural environments in which the moral individuality of the average man is developed*".[16]

In the Gramscian theoretical framework of hegemony, this unquestioned worldview – due to its commonsensical nature, is elaborated by the elites by assuring that their ideas, interests and values are weaved deeply into the social and cultural fabric of society, through the use of its own prestigious philosophers, artists, politics and similar actors as the carriers (Adamson 1980; Morton 2007). In the end, the paradigm popularised by these actors becomes "uncontested common sense" by all segments of society (Birchfield 1999).

In sum, we must talk of hegemony as the construction by elites of a reality that reinforces the status quo, by developing a commonsensical worldview in the peripheral sectors of the population that legitimizes and validates unfairness and exploitation, by motivating them to think and act in structures and terms that are beneficial to the elite (Morton 2007).

Sustainability came as a response to the conclusion reached by Limits to Growth, that unbridled and permanent economic growth would inevitably lead to systemic collapse. By arguing that technological development would help us push these limits further and further, or even eliminate them, sustainability enabled by discursive means, to continue the course in the face of the facts (the effects of which, 50 years later, we are now suffering).

This is why it could be argued that sustainable development is hegemonic: it reinforces the idea that economic growth and market are "natural" and "unavoidable", and now ever-expanding and never-ending, because we have found the way to make it "sustainable". Sustainability thus constitutes an ideal hegemonic concept, as it is attractive to everyone, everywhere (who doesn't want to live in a better, greener world, enjoying increasing states of affluence?) and has been already embraced by most countries the world over, thus spontaneously consenting and

[15] Gramsci (2007), p. 12.

[16] Gramsci (2007), p. 419.

accepting (because supposedly evident and commonsensical) all the ideological content that has been embedded therein (Escoto 2009).

Under this theoretical political ecology framework, we could also argue that sustainability is a successful hegemonic discourse, in that "a*s more and more professionals and activists adopt the grammar of sustainable development, the reinvention of production conditions produced by this discourse will be more effective. Institutions will continue to re/produce the world as seen by those who rule it*".[17]

Consequently, if sustainable development is indeed a hegemonic discourse, then it represents Northern interests, seeking to "*control the resources of others in the name of planetary health and sustainability*",[18] leading them into a determined development path, agreed by all parties to be normal and natural, and preventing any deviation thereto by being automatically judged as counter-intuitive or dangerous (Gaventa 2006).

As a token of how engraved the discourse of sustainability is in the current *zeitgeist*, if a nation were to contest or reject sustainable development to pursue a different path, in first place it would be rebuffed as an act of environmental, social and economic backwardness and selfishness. However, this is just a mere hypothetical scenario, as currently such a move is not even considered or discussed, as it would seem nowadays as counterintuitive, illogical – indeed, "un-natural" as it is part now of the official and unofficial language and psyche.

And so, the band has played on under the umbrella of sustainability, but things have not necessarily improved. Quite to the contrary, as the state of our world has shown by now, we are heading straight the other way.

6 Where to from Here?

"*Lawyers, I suppose, were children once*" -Charles Lamb.

With this epigraph, Harper Lee (2005) begins her masterpiece on racial inequality in the United States, and I often reflect on it, when writing about sustainability from my microcosmical perspective, which is that of a lawyer that has spent most of his life trying to understand what sustainability actually means.

After 25 years as a practitioner, I have studied, read, written, discussed and researched my fair share of this subject. I have been a firsthand witness of the progress and setbacks achieved in the evolution of the discourse, with the new technologies and strategies to achieve a greener, more just world.

And yet, as the years have rolled by, I have also been how, despite all the efforts, all the summits worldwide, panels, technologies, international treaties and re-packaging of the discussion, we are still increasingly harming our planet, by putting economy before ecology.

My issue thus with sustainability, is that it has become such an integral part of our culture and way of life, that in our acritical embrace of the concept, we have also

[17] Escobar (1996), p. 64.

[18] Harvey, quoted in Peet and Watts (1996), p. 10.

maintained the fetish for economic growth that has led us to the current state of affairs, where ecosystemic and climatic collapse are in the order of the day. I believe that sustainability has served as an intellectual analgesic, removing the pain of looking at the cold hard facts and taking the difficult decisions to change our economic and developmental models, and instead making us think that somehow, someway, things are being taken care of and will turn alright in the end, since our economies are getting "greener" and "decarbonized"—"sustainable".

However, as Nicholas Georgescu-Roegen (1971) argued -and, to some extent, was successful in proving- throughout his academic career, thermodynamically speaking, our current economic model, geared toward perennial growth is physically impossible and, in fact, unsustainable.

Hence, inasmuch as our economic model keeps fixated on growth, sustainability is just an enabling concept that creates the fiction that a perpetual growing level of consumption is possible. Ironically, our acritical embrace of sustainability is ushering in what the concept is trying to prevent: that future generations struggle to survive with the ecosystems and natural resources that they will inherit. Indeed, like Count Ugolino in Dante's Inferno, we are "sustainably" gnawing at our own limbs and eating our children, in order to keep on growing.

Notwithstanding everything said to this moment, I believe it important, nevertheless, to introduce a caveat to my critique. Sustainable Development Sustainability for sure was a step in the right direction, and making the world pay attention to the way we live, consume and grow is undoubtedly more valuable than paying no attention at all.

Furthermore, the legal grounding that we have created around sustainability has been quite effecting, in that it modulates, restraints and guides human behavior in its relationship to the environment and society.

However, the fact remains that we are pushing the consequences and limits further ahead into the future, but they do exist, and we must acknowledge them and act in consequence. My core point is that sustainability is preventing us from having this discussion, which needs to happen, be fostered and magnified.

This is why, precisely, in my view Lamb's quote becomes so urgent and relevant to this day: children in their pursuit of understanding and construction of their world, approach knowledge without preconceptions and assumptions. They discard what is unconvincing and are pliable to new points of view when presented with new and more convincing information.

Sustainability needs not to be a monolithic concept, nor be considered as a rigid, static and final model in our social progression. As lawyers, we need to keep pushing forward the narrative, to question the model and argue against a state of things that, in the face of evidence, does not work.

If it is true that the law is at the same time, codification of power and a stalwart against it, the sustainability lawyers hold a prime position to make the concept evolve into something else, something new, that is effective in holding power accountable and servient to the people -the ones that exist now and the ones that are to come. This entails the duty of challenging the notion that economic growth is necessary and inevitable.

Lawyers are instrumental in arguing and providing the groundwork for a better, more just world. One where we finally realize that ecology is the basis of our economy, and not the other way around. To me, this is what being a sustainability lawyer really means.

Make no mistake, this is the stuff that lawyers are made for.

References

Adamson W (1980) Hegemony and revolution: a study of Antonio Gramsci's political and cultural theory. University of California Press, London

Biel R (2007) El Nuevo Imperialismo: Crisis y Contradicciones en las Relaciones Norte-Sur. Siglo XXI, Mexico City

Birchfield V (1999) Contesting the hegemony of market ideology: Gramsci's 'Good Sense' and Polanyi's 'Double Movement'. Rev Int Polit Econ 6(1) Spring: 27–54

Club of Rome (2024) About us. https://www.clubofrome.org/about-us/

Daly H (2007) Ecological economics and sustainable development, selected essays. Edward Elgar, Cheltenham

Dresner S (2008) The principles of sustainability. Earthscan, London

Dryzek J (2005) The politics of the earth. Earthscan, Oxford University Press, Oxford

Escobar A (1996) Constructing nature. Elements for a post-structural political ecology. In: Peet R, Watts M (eds) Liberation ecologies, environment, development, social movements. Routledge, London, pp 46–68

Escoto C (2009) The sustainability of capitalism: a Gramsci an analysis of the sustainable development discourse. Dissertation, University College, London

Gaventa J (2006) Finding the space for change: a power analysis. Inst Dev Stud Bull 37(6):23–33

Georgescu-Roegen N (1971) The entropy law and the economic process. Harvard University Press, Cambridge

Global Forest Watch (2024) Global Deforestation Rates & Statistics By Country. https://www.globalforestwatch.org/dashboards/global/?lang=en.

Gramsci A (2007) Selections from the Prison notebooks. Lawrence & Wishart, London

Lee H (2005) To kill a mockingbird. HarperCollins, New York

McCarthy J (2016) The most and least polluted rivers and waterways in the world. Global Citizen. https://www.globalcitizen.org/es/content/clean-dirty rivers-water-polluted/

Meadows D, Meadows D, Randers J, Behrens W III (1972) Limits to growth. Earth Island, London

Morton A (2007) Unravelling Gramsci, Hegemony and passive revolution in the global economy. Pluto Press, London

National Geographic (2023) Plastic bag found at the bottom of World's Deepest Ocean Trench. https://education.nationalgeographic.org/resource/plastic-bag-found-bottom-worlds-deepest-ocean-trench/6th-grade/

Peet R, Watts M (1996) Liberation ecology. Development, sustainability and environment in an age of market triumphalism. In: Peet R, Watts M (eds) Liberation ecologies, environment, development, social movements. Routledge, London, pp 1–45

Rist G (2007) Development as a Buzzword. Development in Practice 17(4–5) August: 485–491

Sandbach F (1978) The rise and fall of the limits to growth debate. Soc Stud Sci 8(4, November):495–520

Schmink M, Wood C (1987) The 'political ecology' of Amazonia. In: Little P, Horowitz M (eds) Lands at risk in the third world: local-level perspectives. Westview Press, Colorado, pp 38–57

Skillington T (1997) Politics and the struggle to define: a discourse analysis of the framing strategies of competing actors in a 'New' participatory forum. Br J Sociol 48(3):493–513

World Commission on Environment and Development (1987) Our common future. Oxford University Press, Oxford

World Meteorological Organization (2023) Greenhouse Gas concentrations hit record high. Again. https://wmo.int/news/media-centre/greenhouse-gas-concentrations-hit-record-high-again
World Wildlife Fund (2024) How many species are we losing? https://wwf.panda.org/discover/our_focus/biodiversity/biodiversity/

Carlos Escoto Carranza co-leads the Environmental practice and is the coordinator of the Environmental, Social and Governance (ESG) practice at Galicia. He is an expert in environmental law and brings sophisticated regulatory and sustainability insight to the most complex transactions and projects in his country. With over 22 years of experience, he is a recognized specialist in the development of infrastructure, renewable energies, carbon markets and nature-based solution projects, oil & gas, mining, manufacturing industries, and tourism projects. Carlos obtained his Law degree in 2007 from the Instituto Tecnológico Autónomo de México (ITAM) and holds an MSc in Environment and Sustainable Development by University College London (UCL, 2009), where he studied as a British Chevening Scholar. In addition, he is a LEAD Fellow, after completing his specialisation studies at the Colegio de Mexico (COLMEX) as part of the XVIth cohort of this international program in 2011. When his two small children allow it, Carlos likes to go scuba diving in the cenotes and caves of Mexico and likes to listen to old Beta Band records.

Drop by Drop: Resiliency and the Pursuit of Personal Ideals and Professional Passions

Shenandoah Johns

Abstract

This chapter explores the meaning of sustainability leadership, and how fulfilling it can be to weave personal experiences and passions into a sustainability career. It also highlights the contrasting challenges facing sustainability professionals today. Tensions—urgency vs. patience; ideals vs. reality; success vs. winning; and mandatory vs. voluntary—create a platform requiring leaders to be adaptive and creative.

1 Introduction: A Reluctant Lawyer

I am a reluctant lawyer propelled forward by my stubborn Mennonite ancestry.

My four siblings and I were raised to believe that hard work and perseverance could overcome any obstacle. And while our father, a teacher, accepted nothing less than excellence in academics, he ensured that we could differentiate idealism from pragmatism. Like my siblings, I was not allowed to get my driver's license until I could both change a flat tire and do an oil change. Weekdays were for studies; Saturday was for housework and meal prep; and Sunday was the day of rest.

In semi-arid southern Saskatchewan, the summers were windy and dusty, farmers discussed drought and water was precious. Long winters bred a love of reading, and all other spare time was spent at the hockey arena perfecting crossovers, squeezing every dollar of value out of the family's season pass.

My mother was diagnosed with breast cancer when she was only 31, and the five of us were motherless two years later. I was just twelve years old. In fact, within just a span of a few years, our small town of 1000 lost a number of young people to

S. Johns (✉)
West Fraser, Vancouver, BC, Canada
e-mail: shenandoah.johns@westfraser.com

A. McLaughlin, C. Chell (eds.), *Building the Case: How Lawyers are Leading on Sustainability*, https://doi.org/10.1007/978-3-031-88455-9_14

cancer. My mother was the first of two women who left five young children behind. One family lost three generations in little more than a year—first a well-known farmer, followed by his adult son, and then the son's 15-year-old daughter. Even then I wondered, what was happening in our town? What environmental issue was causing this loss of life?

It was my mother that had wanted a houseful of children. After she died, my father survived by running a tight ship. Television was a strictly monitored treat. We were limited to one hour of screen time a week unless we watched Hockey Night in Canada with our dad, in which case we could watch the entire game. I daresay this strict upbringing bred some cheekiness in me. My dad watched Matlock voraciously; one day he caught me peeking around the corner to sneak some screen time, so I blurted out that I wanted to be a lawyer and that I was just doing research. And at the tender age of 12, I learned what a person's word meant to my father.

I tried to walk back that unintended commitment many times over the years. I thought I would be better suited to looking after people. I begged to go into nursing. I made a case for following in my dad's footsteps and becoming a teacher. Most of all I just wanted to be a mom and be there for my kids. All these things I felt were more noble than the legal profession.

Instead, I entered law school because of a gamble I lost with my father. I was 19, without an iota of confidence. He bet I would get in; I thought I wouldn't be admitted because I was too young. The cost of that lost bet? I baked him 100 apple pies and I dove into a career I didn't have the heart for. Years later, I learned that my dad felt motivated to ensure his oldest child had the financial stability to look after the four younger siblings in case anything happened to him, and so, it was easy to forgive his dogmatism when it came to my career path. Coming around on the career itself took a bit longer. Three decades later, I've grown to love and admire my profession and now understand how it has equipped me to bring value and impact where I work.

As I fought the trepidation of law school, stubbornness and my ancestry kicked in. I had a criminal law professor who told us that one in three of the first years would not continue with a career in law. Another professor told us that 50 per cent of women would quit in the first five years of practice. I became someone who was absolutely committed to avoiding the wrong (quitter) side of those statistics.

I wanted to be a woman that changed history and left the world a better place. For a minute, I wanted to be an environmental lawyer and—before Erin Brockovich was a household name—I wanted to figure out what happened in my town that led to so many deaths, including my mom's. Although there was never an inquiry into whether the deaths in my hometown were connected through environmental factors, my resolve to be a problem solver in my circle of influence was fierce and unshakeable.

2 The Skills That Mattered

When I started law school, I was a naïve 19-year-old. I thought that law was about justice. I have had a rich career, filled with experiences that have shaped me and my community. But I've also learned just how elusive my ideals were.

Early in my career, I worked on residential school litigation. To this day, I maintain it was the most important work I ever did. Not because I transformed anyone's life, though I wish I could say I did. It was so important because I was fortunate to be early in my career when I learned to listen.

As a junior lawyer, my job was to interview the fragile people who walked through our doors as they described the atrocities that happened to them when they were ripped away from their families and placed in schools, sometimes hundreds of miles away.

They were taught in another language with a curriculum that was based on a different culture. Imagine a government-endorsed school of evil, and then fill it with society's most vulnerable: children. The cruelty visited on Indigenous people in residential schools became a wicked, wicked problem that is as tragic as any of the atrocities visited on any people in our history.

Quickly my naivete fell away, as it was immediately clear that no financial settlement would ever be enough to remedy the multigenerational trauma and cultural scarring. I was a young, white woman. We weren't wealthy, but we never grew up wanting and love abounded.

I barely had life experience, let alone wisdom, so all I could do was listen. That lesson was more important than any of my law school classes. The two best tools I had now learned by 22 were to listen and to work harder than anyone I know.

3 Finding My North Star

Another important lesson I have learned is that you can't teach people to care. But perhaps the inverse is also true.

About 20 years after graduating from law school, I sat in the office of a successful Vice President at the company I worked for at the time, and she gave me the worst career advice I've ever received: "You need to care less."

She was trying to help me get ahead when she told me that I cared too much about others and that I needed to be more selfish. She was referring to how hard I worked and how much I cared for the people on my team. I sat back in my chair, unable to reply. She was in incredible shape, drove a Porsche, had immaculate style and was breaking glass ceilings. On the other hand, at this point between work and my home life, I was feeling stretched thin.

I had always wanted a big family. My family is everything to me, and my siblings each have superpowers I look up to and admire. I married a man with two daughters from his first marriage, and then we had another 4 children. I always worked, and for many years I was our family's main breadwinner. After 5 years or so in private practice, I had moved into an in-house role. I loved having only one client, and it gave me a chance to help a business succeed. After spending a decade in the law department, I stepped out into a series of business roles. But I was tired. Kraft Dinner and soup was practically a gourmet supper with my kids.

I was determined to make up for all my shortcomings by working harder than anyone around me. The team I led was doing amazing work, but I had a thankless leader. My colleague's solution: care less. I left her office thinking I had lots to learn about success, and I needed to think hard about transforming my approach. But something didn't sit right in my heart. In fact, her advice bothered me so much that I had a few sleepless nights, and I worked through my next greatest lesson.

As I challenged myself, I decided I never, ever wanted a job where I didn't care passionately about the work I was doing. A measure of my worth had to be that I would only work on things I cared about with my whole heart. I vowed to have a job where I could work on things that lined up with my values, alongside people I loved. Integrity was table stakes, and empathy, excellence and generosity were those characteristics that defined my brand.

If I cared too much about my work, and that cost me career advancement, then that was a cost I was prepared to accept—not reluctantly, but joyfully.

4 Determining My Non-Negotiables

I've concluded that the reality of living with integrity is there is often a trade-off—a reality that so many want to ignore.

Wherever I have worked, I have been hailed by women and men alike as evidence that "you can have it all": 2+4 kids, a 25-year marriage and a successful career. I wish I could say that living your values will only lead to wild success. An inconvenient truth is that living your values can come at a steep cost.

I am a huge sports fan and, wherever possible, I love cheering for Canadian teams. When it comes to baseball, that means the Toronto Blue Jays. I still remember, during my first year of law school, watching the '93 World Series game from my apartment when Joe Carter hit a three-run walk-off home run to win the crown.

Decades later, the Jays did not have a particularly good 2017, but they did have a pitcher who was making a name for himself as a closer. Roberto Osuna was named as an All-Star and the following April, he became the youngest pitcher in Major League Baseball (MLB) history to record 100 saves. But just a month later, in May 2018—amid his pitching prowess—he was charged with the vicious assault of his domestic partner, Alejandra Roman Cota.

MLB conducted its own investigation, and based on the evidence they reviewed, Osuna was given a 75-game suspension for violation of the MLB Domestic Violence Policy. The criminal charges were dropped because Alejandra returned to Mexico and refused to come back to Toronto to testify against Osuna.

The Toronto Blue Jays had a zero-tolerance policy to domestic violence and in July 2018, Osuna was traded to the Houston Astros. The Jays missed the playoffs in 2018 and 2019. On the other hand, Houston made the playoffs in 2018, and in 2019 they won the American League Championship against the New York Yankees, clinching their spot in the World Series. The night of the win against the Yankees, an executive with the Houston Astros, Brandon Taubman, turned to reporters following the team – including three female reporters, one of whom was wearing a purple domestic violence-awareness bracelet – and said, "Thank God we got Osuna! I'm so f------ glad we got Osuna!"

There isn't going to be a Disney movie about that ending. Living their values didn't translate into postseason success for the Toronto Blue Jays.

I have faced more than one choice at work where doing the right thing has come at a cost. My generation pokes fun at youth today, where there are medals and awards passed around for just having a pulse and showing up. Participation trophies. Ironically, this same generation of mine coined the phrase "career-limiting move", recognizing that pointing out that the emperor has no clothes or speaking truth to power can have devastating consequences.

I simultaneously war against the reality of injustice and unfairness, and I am at peace with refusing to negotiate on the things that matter most to me. I will fiercely advocate for women and the underserved in our society, and I will relentlessly advocate that environmental excellence is as important as anything else for those companies seeking operational excellence. These messages are rarely comfortable conversations.

5 Playing the Long Game

One of the early, ironic lessons I learned from a very senior lawyer is that there is such a thing as winning too big.

I often heard this man crow about how he and the business team outsmarted and out-negotiated another corporation and long-term partner to secure financial terms beyond his wildest expectations.

I was the lawyer who inherited providing advice over the remaining term of the contract. Having secured such "great" terms, we found ourselves in constant disputes and endless renegotiation. It was time-consuming and expensive. More importantly, though, this other corporation was not incented to perform well under the contract. As a result, our operational performance suffered, and the cost to the company I worked at was significant. "Winning" the commercial negotiation came at an unintended price for the company, costing management commercial relationships and costing shareholders money.

Working in the field of sustainability was an opportunity for me to delve into issues I was passionate about and combine that passion with driving business value over the long term. I have had an opportunity to build trust in businesses that I hope will thrive well beyond the conclusion of my career. There is deep satisfaction in that kind of work.

I recently had a conversation with my children about my desire to secure their future by working hard on things that mattered, like human capital, climate change and biodiversity loss. We talked about what equity means, and how far we still must go as a society for social justice. We talked about what environmental justice means, and what the right to a healthy environment should look like. And while all these things are noble, the reality is that I am not working to add even more value for future generations, I am working to reduce the debt our society is leaving them with.

I try to be a realist and bring pragmatism to whatever I do. Some think I am cynical, but I think it's different. I am an idealist, but my DNA is wired to drive results. I love that most of the sustainability experts I work with are tackling issues that they care about very much.

Often, I meet subject matter experts who know way more than I do about particular sustainability issues, which tend to be technical and complex. I think of myself as a technician working in the "Tricky Issues Lab". The opportunity, from a personal development perspective, is the mash-up of idealism and pragmatism that unleashes creativity and brings collaboration and innovation to the table in order to problem-solve wicked issues. The "lab techs" I work alongside are empowered to create solutions in a really special way.

I think that the magic formula for sustainability leaders is to become a master translator; taking sustainability theory and academia, and molding and implementing them into actions that everyone understands and cares about.

When I look at the little things that show up in the lives of my children, I see how my oldest daughter, a chef, prepares food and hands it out to unhoused people on the way to and from her workplace. My second daughter prefers to thrift or make her own clothes, and she is adamantly opposed to the waste of fast fashion. My youngest daughter battles depression. She fiercely canoed the Yukon River, stopping at Indigenous communities, learning about their history, the impacts of climate change on their lives and the fragile ecosystem they live in. At 18, her perspective on people and their value shows up in her artwork. And as I write this, my son is working at a manatee conservation sanctuary in Central America.

Success then looks at not the recognition of a few, but rather of how individuals across society and companies demonstrate care and commitment through everyday actions.

6 A Union of Urgency and Patience

For any company aspiring to embed sustainable practices into its strategy and build resiliency to navigate a world of increasing volatility, it should be recognized that this transformation will span multiple leadership changes and it requires senior-level buy-in throughout.

A sustainability culture cannot easily be *taught*—it needs to be *caught* and woven into the fabric of how things are done.

So, on one hand, every sustainability professional I know brings passion and urgency to their work. And yet, surviving the highs and lows of passion requires unending patience and an openness to learn from setbacks and failure. The juxtaposition of urgency and patience is challenging to navigate.

Experts in organizational effectiveness say it can take several years to navigate each step of the progression toward operational excellence, as illustrated in Fig. 1.

Fig. 1 Sustainability leadership

In fact, moving from a culture that is regressive to a culture where a company is committed to positive impacts across multiple facets can take decades.

By losing a bet with my dad, I made a lot of apple pies. My first pie was not very good. The crust was soggy, and the apples were not cooked through. Hundreds of pies in, and 30 years later, I can bake a mean apple pie—most of the time. But every once in a while, especially when I'm looking to experiment and shake things up, I can turn out something that makes better compost than dessert and I need to start again.

In the field of sustainability, the stakes are much higher than apple pie—but the point around resiliency remains. In our profession, we often feel like when we take one step forward on an issue, we take two steps back on another. Except here we're talking about our planet; Mother Earth is hardy, yet fragile, just like many of her inhabitants. The stakes are safety, health and well-being—for our coworkers, our families, our communities and our environment, today and for future generations. Perseverance towards progress—even amidst setbacks that can be very deeply felt—is critical.

7 I am a Lawyer Who Chooses to Lead

To this point in my career, I have loved working "in-house". I have been able to contribute to the business strategy of some great companies and bring pragmatic solutions to a profession that, at times, is accused of being overly technical and introducing more challenges than it solves.

To be both practical and effective, one must become very good at managing risk. However, I am becoming more and more concerned about certain legal perspectives that are showing up in sustainability discussions. The regulation of sustainability practices that have been part of leadership conversations for decades (climate, nature, materiality, diversity) often brings highly risk-averse, conservative views to the Tricky Issues Lab. This stifles creativity, and in my view, might threaten the cost, and ultimately the existence of, the Lab.

My training as a lawyer helps me understand this risk conversation, but suddenly the predominant investment of companies has become "compliance". As the corporate world re-centers around these obligations, rather than understanding and encouraging voluntary leadership action, I have seen an abandonment of operational excellence strategies that are truly visionary and inspirational (see Fig. 1) in exchange for doing only what the law requires. I believe it is imperative that aspiring sustainability leaders do not set compliance as their vision. In many ways, managing to compliance is unrewarded risk management from an investor's point of view. Though this is a very natural place for lawyers to practice, true leaders recognize that real risk management goes beyond compliance and includes taking on the right risks to generate incredible business value.

When I look at just a few of the escalating impacts in society today, like market and geo-political volatility and climate change, I do not think a risk-managed company is necessarily a sustainable one. Proactive practices, including extending your influence beyond the traditional areas of corporate control, will become necessary to ensure the long-term resilience of corporate strategy.

As a company moves to protect its reputation and shareholders from allegations of "greenwashing", it often adopts behaviours and practices that diminish its voluntary leadership efforts. In contrast to greenwashing, this is called "greenhushing", and my view is this will inhibit operational excellence and the practice of embedding sustainability practices at all levels of corporate culture.

I am concerned that instead of raising the floor, sustainability overregulation will lower the ceiling, and we will abandon the vision of "net positive" companies in acceptance of risk management. The upside for the legal profession is that we will need more lawyers. But specifically, we need lawyers who care. This is what is required if we want to transform corporate strategies from near-term investor returns to consider long-term impacts that promote resiliency and create companies that thrive through various scenarios.

8 Conclusion: A Single Bead of Water

In my closet, I keep a stack of copies of a beautiful book by Michael Nicoll Yahgulanaas called *Flight of the Hummingbird.*

From time to time, as I recognize the sometimes big, sometimes small, efforts of the people around me, I will give them a copy as encouragement. The book is a parable for the environment, and it has origins with the Quechan people of South America and the Haida people of the North Pacific.

In the story, a terrible fire is consuming the Great Forest. While all the animals flee, the hummingbird—a symbol of wisdom and courage—picks up water and, bead by bead, drops it on the raging fire. As all the animals warn her of the danger, the hummingbird replies, "I am doing what I can."

I am deeply grateful for my upbringing; the intersection of my ideals of leaving the world better than I found it; and the pragmatism of doing what I can.

Shenandoah was born in Whitehorse, Yukon Territory, and has made homes in Saskatchewan, Alberta and British Columbia. Her father, who retired to a cherry orchard, earnestly used all of his resources to give his children the best he could, and knows that he raised children (a lawyer, nurse, engineer, cabinet maker and teacher) who care very much about leaving the world better than they found it.

Insatiable curiosity, a love of change, solving tricky problems and possibly gypsy blood have caused Shenandoah to take on eclectic career challenges. After leading a legal team, she was asked to create a Capital Planning and Development team that managed $6B of capital, followed by experiences in leading Corporate Strategy, Economics, Environment and Regulatory, and culminating in her current position of Chief Environment and Sustainability Officer at a leading forestry company.

In her spare time, she prefers to choose either reading a good book, escaping to the mountains or nourishing her loved ones with home-cooked food. However, it is as likely that she will be found tackling a mountain of laundry, vacuuming or grocery shopping.

Shenandoah Johns lives in Vancouver, British Columbia with her husband of 25 years, Vincent, and her four children, Sophia, Isobelle, Evangeline and Samuel. Her bonus daughters, Rachel and Rebecca live nearby and are also a vibrant part of her family.

The manufacturer's authorised representative in the EU is Springer Nature Customer Service Centre GmbH, Europaplatz 3, 69115 Heidelberg, Germany. If you have any concerns regarding our products, please contact ProductSafety@springernature.com

Printed and bound by CPI Group (UK) Ltd, Croydon, CR0 4YY
01/10/2026
02225850-0001